SCIENTIFIC PRACTICES AND
THE PORTUGUESE EXPANSION IN ASIA (1498-1759)

HISTORY OF MATHEMATICAL SCIENCES: PORTUGAL AND EAST ASIA II

University of Macau, China 10 – 12 October 1998

edited by

LUÍS SARAIVA
University of Lisbon, Portugal

World Scientific

NEW JERSEY • LONDON • SINGAPORE • BEIJING • SHANGHAI • HONG KONG • TAIPEI • CHENNAI

SCIENTIFIC PRACTICES AND
THE PORTUGUESE EXPANSION IN ASIA (1498-1759)

HISTORY OF MATHEMATICAL SCIENCES: PORTUGAL AND EAST ASIA II

Scientific Committee:

Jean Dhombres (CNRS: Paris, France)
Catherine Jami (CNRS: Paris, France)
Liu Dun (Chinese Academy of Sciences: Beijing, China)
Luís Saraiva (Coordinator of the Committee; CMAF: Lisbon, Portugal)

Local Organizing Committee:

Raymond Cheng (Chairman: FST; University of Macau: China)
Isabel Loureiro (Co-Chairman: FST; University of Macau: China)
Denise Sam (FST; University of Macau: China)

Cover design: Claudia Hora, from a painting of Beijing Observatory, published in *Revista de Cultura*, 21, 1994, Instituto Cultural de Macau.

Composition: Carlos Perpétuo

FUNDAÇÃO CALOUSTE GULBENKIAN

spm SOC. PORTUGUESA de MATEMÁTICA

CMAF UL

This proceedings is made possible by a joint grant from the Calouste Gulbenkian Foundation, the Portuguese Society of Mathematics (SPM) and the Center for Mathematics and Fundamental Applications (CMAF).

Published by

World Scientific Publishing Co. Pte. Ltd.

5 Toh Tuck Link, Singapore 596224

USA office: 27 Warren Street, Suite 401-402, Hackensack, NJ 07601

UK office: 57 Shelton Street, Covent Garden, London WC2H 9HE

British Library Cataloguing-in-Publication Data
A catalogue record for this book is available from the British Library.

HISTORY OF MATHEMATICAL SCIENCES: PORTUGAL AND EAST ASIA II — SCIENTIFIC PRACTICES AND THE PORTUGUESE EXPANSION IN ASIA (1498–1759)

ISBN 981-256-078-5

Printed in Singapore by World Scientific Printers (S) Pte Ltd

ACKOWLEDGEMENTS

We would like to express our gratitude to all who made this Conference, and the publication of its Proceedings, possible.

First of all we would like to thank the speakers, for their talks, for their availability to do research on the themes of the Meeting, and for their commitment all the way through this process, from the early stages of preparation of the Meeting, to the interest and enthusiasm they showed during the Conference, and to the publication of these Proceedings. In particular, I wish to thank Professor Catherine JAMI, who played an essential part in shaping this Conference, and gave helpful comments for the revision of the papers.

Secondly I would like to thank the organisers of the Meeting, the University of Macao and Centro de Matemática e Aplicações Fundamentais, and all its sponsors: Câmara Municipal das Ilhas, Comissão Nacional para as Comemorações dos Descobrimentos Portugueses, Direcção dos Serviços de Turismo de Macau, Fundação Macau, Fundação Oriente, Instituto Cultural de Macau, Instituto de Cooperação Científica e Tecnológica Internacional, Instituto Politécnico de Macau, Instituto Português do Oriente, and Leal Senado de Macau, for making this Conference possible.

I also wish to thank the Calouste Gulbenkian Foundation, an essential institutional reference concerning culture in Portugal, which is sponsoring the publication of the Proceedings, and Sociedade Portuguesa de Matemática, who made this sponsorship possible.

Last, but not least, I want to thank the members of the local organising committee, chaired by Doctor Raymond CHENG, and the many scholars and students who helped the Conference to run so smoothly.

Luís SARAIVA
Centro de Matemática e Aplicações Fundamentais / Universidade de Lisboa
Lisbon, December 2001

FOREWORD

Luís SARAIVA
CMAF, University of Lisbon

The Conference *"History of Mathematical Sciences: Portugal and East Asia II"* took place at the Faculty of Science and Technology of the University of Macao from the 10th to the 12th of October 1998. It was organised by the University of Macao and Centro de Matemática e Aplicações Fundamentais, and was also sponsored by Câmara Municipal das Ilhas, Comissão Nacional para as Comemorações dos Descobrimentos Portugueses, Direcção dos Serviços de Turismo de Macau, Fundação Macau, Fundação Oriente, Instituto Cultural de Macau, Instituto de Cooperação Científica e Tecnológica Internacional, Instituto Politécnico de Macau, Instituto Português do Oriente, and Leal Senado de Macau. The simple enumeration of these ten institutions makes clear the interest generated by the Conference and its theme, with the sponsorship coming from some of the most important institutions in Macao and Portugal.

This event began the day after the end of the *First Luso-Chinese Symposium on Nonlinear Evolution Equations and their Applications*, a conference also organised by the University of Macao, which brought together thirty-five researchers, twenty-three from mainland China, three from Macao, and nine from Portugal. It was not a matter of chance that these two events happened in succession. This was done on purpose, firstly to point up the important relationship between practising a science and reflecting on its development, between the making of new knowledge and its critical integration into a historical setting, in short, between the past and the present. Secondly, it showed a willingness to have international collaboration and debate on mathematical issues of the present and the past, recognising that a major factor in the progress of society is the contact between different cultures, backgrounds, experiences, knowledge, ways of thinking and of acting. And thirdly, it clearly showed the University of Macao's willingness to promote this debate and to participate in it. It is fitting that this conference took place in Macao, a meeting point of East and West, a place always known for the coexistence of different cultures.

The Conference *"History of Mathematical Sciences: Portugal and East Asia II"* was intended to be a continuation of the debate started in 1995, at the Convento da Arrábida in Portugal, with the first Conference *"History of Mathematical Sciences: Portugal and East Asia"* [1]. The main aim of these conferences was to analyse the interaction between Europe and East Asia between the 16th and 18th centuries in the field of mathematical sciences, bringing to the fore the role of Portugal as an agent of transmission of European science to East Asia. The 1995 conference, organised by Fundação Oriente, represented a real breakthrough, in the sense that its main purpose was the study of a theme until then almost completely ignored by Portuguese scholars. The relationship between Portugal and East Asia during the 16th, 17th, and 18th centuries had been methodically studied by historians, including Portuguese researchers, in its economic, ideological, cultural, social, and political aspects. However there was a complete different picture when it came to the history of mathematical sciences: few scholars studied this subject, none of them Portuguese. Significantly, the four main histories of Portuguese mathematics (Garção Stockler's in 1819, Rodolfo Guimarães's in 1909, Pedro José da Cunha's in 1929, and Francisco Gomes Teixeira's in 1934) say hardly anything on the subject. Therefore, the Arrábida Meeting, with its three main themes (Mathematical Sciences and Arts: the meeting of East Asia and the European West; The Society of Jesus, agent of transmission of European culture; Portugal and its Asian connections) and, equally as important, the emphasis on the absolute necessity of research done on primary sources (something which has been a recurrent shortcoming in the research work done by the majority of Portuguese researchers on the history of mathematics[2]), was an event that also aimed at stimulating serious research into Portuguese history of mathematics topics.

The holding of this second Conference (and the publication of these Proceedings) also has a special meaning: it is an expression of the willingness to continue the debate started with the Arrábida Meeting,

---

[1] The Proceedings of this Conference were published in 2000 by Fundação Oriente.

[2] This remark does not concern the history of nautical sciences, which blossomed in the 20th century, due to the works of researchers such as Joaquim Bensaúde (1859-1952), Luciano Pereira da Silva (1864-1926), Abel Fontoura da Costa (1869-1934), Luís de Albuquerque (1917-1992), and others, and has a history of its own.

the corroboration that there are both researchers interested in studying and exchanging views on these issues, and a public keen to know about them. It is now clear to us that there are conditions to organise periodic conferences on this general theme, exactly as the organisers of the First Conference intended, when the idea of these meetings initially took shape.

The limiting dates chosen for this second Conference are the arrival of Vasco da Gama in India in 1498 and the expulsion of the Society of Jesus from Portugal in 1759. The landing of Vasco da Gama in India in 1498 opened a new era in the relations between Europe and Asia, in particular regarding the circulation of scientific knowledge. This led to major social and intellectual changes in both continents. The Society of Jesus controlled education in Portugal and in the Empire in the middle of the 16th century. It was therefore central to the network of knowledge transmission until the Society was expelled from Portugal in 1759.

There were three main themes for the Conference, which were described as follows in the Conference's program:

1. ***Pacing and Mapping the World*** — First and foremost, the newly explored domain had to be mapped: the development of cartography, the collection of astronomical observations, and of data relating to nature and to societies encountered, resulted in a new picture of the world. European maps began to use coordinates and different kinds of mathematical projections. These transformed human-localised space into a uniform system of reference. To Jesuit missionaries, maps, which conveyed many Christian symbols, epitomised Western experience. The encounter between this system and those used until then in Asia sometimes resulted in interesting syntheses.

2. ***Transmission, Translation and Teaching*** — Translation into hitherto unknown languages in many cases turned out to be a prerequisite to teaching the Gospel and those aspects of European science that were linked to it in the missionaries' minds. The reception of these elements of scientific knowledge by local elites can only be understood in the light of local systems of learning. In the other direction, it was knowledge about Asia rather than Asian knowledge that was brought back to Europe. Although some transfers of technology did occur, little of Asian learning and techniques was deemed of interest by Europeans, including the Jesuits themselves.

3. ***Science as a Political Tool: the impact on Asian Societies*** — Ambassadors and missionaries were among those who used science as a means of increasing their influence in Asia. Instruments were given as diplomatic gifts, Jesuits served as astronomers in various countries. This resulted in the integration and partial re-creation of European science in different cultures and societies.

The organisers did their best in this Meeting to present a representative selection of researchers on the theme of the Conference. Of a total of fifteen speakers, four were from mainland China, one from Macao, one from India, two from Japan, one from Korea, and two from France, three from Portugal, and one from Spain. It is worth pointing out that, in contrast to the Arrábida Meeting, there was a majority of Asian scholars[3]; and that for the first time there were researchers from Japan, India, Korea, and Spain. The organisers tried to have scholars from Macao and Hong Kong Universities at this Conference, as they felt their contribution would be relevant to the theme of this Meeting. At one point before the Conference it was thought that this aim would be fully achieved, as there was one researcher from each University in the list of speakers. As it happened, FUNG Kam-Wing (Hong Kong University), initially on the speakers list, in the end could not participate in the Conference. Only FOK Kai Cheong (University of Macao) was able to present a talk at this event, but his contribution, "Science and technology and the change of the Portuguese image in Ming and Qing China", could not be included in the Proceedings.[4]

During the Conference the talks were grouped into themes. In these Proceedings this structure is maintained[5]. We include here twelve of the fifteen texts presented at the Meeting. The first paper is from the ***opening talk***, by Luís Filipe BARRETO (Faculdade de Letras, University of Lisbon; Portugal), on "Macau: An Intercultural Frontier in the Ming Period". This is followed by a series of papers on ***cartography***: "Survey and study of pre-1900 Chinese maps seen in Europe" by LI Xiaocong (Beijing University; China), "Western knowledge of Geography reflected in Juan Cobo's *Shilu* (1593)" by LIU Dun (Institute for the History of Natural Science, Chinese Academy

---

[3] For the Arrábida Meeting, we had a total of fourteen speakers. Two were from China, one from the United Kingdom, three from France, one from the Netherlands, one from Italy, five from Portugal, and one from the U.S.A.

[4] For an analysis of a related subject, see Fok Kai Cheong, *Estudos sobre a instalação dos Portugueses em Macau*, Gradiva, Lisboa, 1996.

[5] The titles of some of the papers differ slightly from those listed in the Programme of the Conference. This is because of changes made by the authors to their papers.

of Sciences; Beijing, China), and "The continuing influence of the Portuguese: the contribution of *A New Interpretation of World Geography*" by WANG Qianjin (Institute for the History of Natural Science, Chinese Academy of Sciences; Beijing, China). Then we have a section on ***The Jesuits in East Asia***. Illness prevented Professor Ugo BALDINI (Università degli Studi G. D'Annunzio; Rome, Italy) from participating in this Conference and presenting a talk on this topic; hence here we have a single paper, by Catherine JAMI (Centre National de la Recherche Scientifique; Paris, France): "Teachers of Mathematics in China: the Jesuits and their Textbooks (1580-1723). This is followed by a section on ***Reciprocal Images***. Here, due to the non-availability of the paper by the scholar from the University of Macao, we have only one paper, by Rui LOUREIRO (Lusófona University; Lagos, Portugal): "News from China in sixteenth century Europe: the Portuguese connection". Next we have a paper on the topic of ***Orientalization and the Construction of Universality***, by Juan GIL (University of Seville; Spain): "The Indianization of Spain in the 16th century". Another paper on this theme, read at the Meeting, "Does science have to be reduced to what is proven through deduction to be universal? A historical question to European science (1498-1759)", by Jean DHOMBRES (CNRS; Paris, France), also could not be included in the Proceedings. The following three papers are on ***Astronomy and Cosmology***. The first two are on China, the third is on Japan. They are "Jesuit Observations and Star-Mappings in Beijing as the transmission of scientific knowledge" by HASHIMOTO Keizo (Kansai University; Osaka, Japan), "The compilation of the *Lixiang kaochenghoubian*, its origin, sources, and social context" by HAN Qi (Institute for the History of Natural Science, Chinese Academy of Sciences; Beijing, China), and "A Japanese reaction to Aristotelian Cosmology", by YOSHIDA Tadashi (Centre for Northeast Asian Studies, Tohoku University; Sendai, Japan). The paper "*Kenkon Bensetsu*: the Japanese translation of the *Treatise of the Sphere* and the reception of Western Cosmology and Astronomy in 17th century Japan" by Henrique LEITÃO (Centro de Física da Matéria Condensada; Lisbon, Portugal) was read at the Conference but could not be included in the Proceedings. Finally, the last two talks were grouped under the general theme ***Other Lands***, and were "Portugal and Korea: obscure connections in the pre-modern sciences before 1900" by PARK Seong-Rae (Hankuk University of Foreign Studies; Korea), and "Translations of Portuguese texts into Konkani and Konkani compositions into Portuguese, with educational influence on literature and art and the transfer of technology" by Joseph VELINKAR (Heras Institute, St. Xavier's College; Bombay, India).

We end this brief introduction by expressing our hope that the publication of the Proceedings of this important Conference will contribute to the dissemination of the contents of the Conference to the general public, to historians, and in particular to historians of science. We hope that in its modest way this book may provide some inspiration for the continuation and deepening of the debate and research on the history of mathematical sciences concerning the complex questions of the transmission of science between different cultures, and in particular that it will help to shed new light on the role of Portugal as an agent of transmission of European science to Southeast Asia. We also hope that it will foster the co-operation of general historians with historians of science, and specially with historians of mathematical sciences. The interdisciplinary collaboration between researchers with different backgrounds and knowledge has always been one of the most fruitful means of advancing research and obtaining new results.

TABLE OF CONTENTS

PICTURE OF CONFERENCE PARTICIPANTS

Front Row, from left to right: Raymond Cheng, Jaime Carvalho e Silva, Catherine Jami, Wang Qianjin.
Back Row, from left to right: Han Qi, Isabel Loureiro, Juan Gil, Henrique Leitão, Hashimoto Keizo, Li Xiaocong, Liu Dun, Yoshida Tadashi, Park Seong-Rae, Jean Dhombres, Joseph Velinkar, Fok Kai Cheong, Luís Saraiva, Alexei Volkov.

PICTURE OF CONFERENCE PARTICIPANTS

Front Row, from left to right: José Francisco Rodrigues, Luís Saraiva, Catherine Jami, Park Seong-Rae, Wang Qianjin, Joseph Velinkar, Jean Dhombres, Raymond Cheng, Alexei Volkov.

Back Row, from left to right: Jaime Carvalho e Silva, Rui Loureiro, Henrique Leitão, Yoshida Tadashi, Liu Dun, Hashimoto Keizo, Li Xiaocong, Han Qi, Vong Seak Weng.

PHOTO OF CONFERENCE PARTICIPANTS

Front Row (from right to left): [illegible] Seong-Kee [illegible] Ubolsri, [illegible] Humphreys, Raymond Cheng, [illegible]

Back Row (from left to right): James [illegible] Kam, Li [illegible], Ho [illegible]

MACAO: AN INTERCULTURAL FRONTIER IN THE MING PERIOD

Luís Filipe BARRETO
Faculdade de Letras, University of Lisbon

This short study is designed to contribute to an understanding of certain aspects of the cultural dimension of Macao between the mid-15th and mid-16th centuries.

First of all, we have to take into account the social conditions during the Ming period that led not only to the establishment of Macao but also to its continued existence as a city with internal and external links between Europeans and Chinese [1].

I

Possibly the first mention of the site of Macao in Portuguese documents is in the geographical treatise *Suma Oriental* written by Tomé Pires between the years 1512 and 1515 in Malacca and Cochin.

Through the maritime and mercantile networks of Southeast Asia and the overseas Chinese communities established there, the Portuguese gathered the information that:

> *"... beyond the port of Quantom, there is another port which is called Oquem at a distance of three days overland or a day and a night by sea; this is the port used by the Lequios and other nations..."* [2].

[1] See Jorge Manuel dos Santos Alves – *Um Porto entre Dois Impérios: Estudos Sobre Macau e as Relações Luso-Chinesas*, Macao, IPOR, 2000.
[2] *A Suma Oriental de Tomé Pires e o Livro de Francisco Rodrigues*, Edição A. Cortesão, Coimbra, I. Universidade, 1978, Book IV, pp. 368-369.

It is possible that the Portuguese *Oquem* is a phonetic transcription of the Cantonese *Haocheng* or the Mandarin *Haojing*, meaning "oyster mirror". Throughout the Ming period, Chinese texts generally refer to Macao as Haojing, as can be seen for example in a letter of 1637 written in Chinese to Father Zeng (the Jesuit priest Álvaro Semedo) from the Chinese Christian community in Haojing/Macao [3].

Around 1515, Tomé Pires transcribed the Cantonese Haocheng as Oquem and stated that in the vicinity of Canton there was a harbour used by the Lequios (from the Kingdom of Ryukyu) and by others he did not identify (mainly the people of Fujian).

The kingdom of the Ryukyu islands, unified in 1429 by Sho Hashi, with trade under royal monopoly, paid tribute to China within the framework of official trade relations, as well as trading under the same conditions with Japan. The Lequios were the main intermediaries in Japanese trade with Southeast Asia, particularly from around 1411, linking Malacca to the southern part of the islands of Kyushu and Tsushima, as well as being involved in the Japanese export trade to China and the flourishing trade with Korea in the 14th and 15th centuries [4].

Forty years later, on 20 November 1555, we have the first text in Portuguese written in Haojing/Macao.

This is a letter written by Brother Fernão Mendes Pinto to the Jesuit Baltasar Dias, the rector of the Jesuit College in Goa:

> *"... today I arrived from Lampacau, the port where we are based, in Macao, which is six leagues further on, where I encountered Father Belchior, who had come here from Canton..."* [5].

Haojing has become Macao, the Portuguese not adopting the phonetic transcription of the Cantonese word for "Oyster Mirror", that is Oquem. The name Macao probably derives from the Fujian phrase *Má Kó Cau*, meaning "Bay of the Ancestral Grandmother". In this bay, also a harbour, there had

---

[3] In Biblioteca Apostolica Vaticana, Bar.Oriente. 154. 1a. On the Chinese resources of the library, see Yu Dong – *Catalogo delle Opere Cinesi Missionariae delle Biblioteca Apostolica Vaticana (XVI-XVIII sec.)*, Vatican City, Biblioteca Apostolica Vaticana, 1996.

[4] See Robert K. Sakai – *De Ryukyu – (Liu-Chiu) Island as a Fief of Satsuma* in J.K. Fairbank, (ed.) – *The Chinese World Order: Traditional China's Foreign Relations*, Cambridge, Massachusetts, Harvard University Press, 1974, pp.112-134.

[5] Letter from Fernão Mendes Pinto dated 20 November 1555 to Baltasar Dias, Rector of the Jesuit College in Goa, in R. Katz and F.M. Rogers – *Cartas de Fernão Mendes Pinto e outros documentos*, Lisbon, Presença, 1993, p.61.

for centuries been a temple to the goddess Má-Kó. A popular goddess along the coasts of Fujian, of Ryukyu and of the south of Japan and Kyushu Island, Má-Kó was the protectress of fishermen, the 'empress of the skies', as well as 'the eye that sees a thousand miles' and 'the ear that follows the wind', harnessing the winds and protecting sailors.

From the Haojing of the Ming-period Chinese to the Macao of the Portuguese from 1555 is not only a question of transcriptions from Cantonese or Fujian.

In Malacca the Portuguese heard of Oquem as a port in the province of Canton, used by the Lequios, and it was the Lequios and the people of Fujian who, from around 1535 at the latest, began using the "Oyster Mirror" harbour regularly and who, in 1555, baptised it Macao from the Fujian phrase for "Bay of the Ancestral Grandmother".

Between 1535 and 1555, the Portuguese started using the port of Haojing regularly in their private and semi-official maritime trading relations between Southeast Asia and the China Seas. These networks of trading partnerships, operating out of Patane and Malacca, linked the Portuguese merchants to overseas Chinese communities in Southeast Asia (known as *Nanyang*) and the intermediaries of Ryukyu, as well as to the maritime trade from the provinces of Fujian and Guangdong and, from around 1542-1543, to the regular trade with southern Japan.

The establishment of the Portuguese in Macao was the result of Portuguese merchants entering into the Asian networks of partnership and competition for the maritime trade between Southeast Asia and southern China and Japan.

The Portuguese presence in Macao arose from groups of 'Asianised' Portuguese or those of mixed blood, that is to say, those who were most closely involved in trade networks and commercial interests within Asia, especially those linking Southeast Asia to the Far East. These networks and interests were far more frequently private or semi-official than state-controlled and official.

Throughout the Ming period, the Portuguese developed Haojing/Macao from an inter-regional harbour to an international port city, linking the seas around southern China and Japan to the Indian and Atlantic Oceans, and even to the Pacific:

> *" ... a port of call for all merchandise coming from India on its way to China and Japan and other places in the Far East..."* [6].

[6] Letter from Ponciano Lanços de Abreu from Macao, dated 4 November 1641, to King D. João IV in Lisbon in Arquivo Histórico Ultramarino, Avulsos, Macao cx I doc.20.

However, the basis of this maritime trade had long existed in "Oyster Mirror"/"Bay of the Ancestral Grandmother". This was what first attracted the Portuguese merchants and provided the conditions that allowed the transformation of Macao during the Ming period into a leading port and one of East Asia's key international maritime cities.

The second social characteristic to be noted, which to some extent is a consequence of the first, is that the importance of the networks and partnerships of Asians and Asianised Portuguese meant that, throughout the Ming period, Macao enjoyed a somewhat indeterminate status, in practice resulting in an informal and unofficial shared sovereignty.

In 1582, an official manuscript report was submitted to the King in Lisbon concerning Portuguese expansion in the seas and along the coasts of Asia. In the anonymous *Livro das Cidades e Fortalezas que a Coroa de Portugal tem no Estado da Índia* (Book of the Cities and Fortresses belonging to the Portuguese Crown in the State of India), it states in relation to Macao that:

> *"... although the land belongs to the King of China, who has his officials there to collect the taxes payable there, they are governed by the laws and administrative systems of the Kingdom of Portugal..."* [7].

Thus, for the central power in Lisbon, the city of Macao was, in 1582, Chinese territory subject to the Chinese fiscal authorities, particularly in respect of payment by the Portuguese of a "*foro do chão*", that is a tax for use of the land, while being at the same time a place governed by Portuguese laws and administrative systems. In other words, Chinese sovereignty was partially shared with the Portuguese.

In 1637, in a letter from the Senate of Macao, we find the following statement:

> *"... this is not our land, conquered by us, like the fortresses of India where we rule... but rather land of the King of China where we do not have an inch of territory, except this city which, while subject to our King, still belongs to the King of China..."* [8].

[7] *Livro das Cidades, e Fortalezas, que a Coroa de Portugal tem nas partes da Índia, e das Capitanias e mais Cargos que nellas ha, e da importância delles* (1582), ed. F. Mendes da Luz, Lisbon, C.E.H.U., 1960, chap.XV, fol.75.

[8] Letter from the Senate of Macao dated 24 December 1637 in A.N.T. Tombo, Documentos Remetidos da Índia, 41, fol.221.

The Senate of Macao is the key institution of this port city — "... *the gentlemen of City Hall are those who rule over Macao...*" [9] — the organised voice of local power in Macao and of its merchant oligarchy.

What was the opinion of this, the highest body representing the Portuguese merchants of Macao? It confirmed Chinese sovereignty over Macao and the situation of partial sharing in which, on Chinese soil, there was a system linked to the Portuguese Crown.

The words written in Lisbon in 1582 and in Macao in 1637, expressing the views of the Portuguese Crown and the local powers in Macao, reaffirm Macao as part of China but including Portuguese elements, the result of their value as intermediaries in international and inter-regional maritime trade.

This is not the place to present a study of the status of Macao during the Ming period. What interests us here is that, in political and social terms, the Portuguese at that time saw Macao as Chinese territory that incorporated a range of Portuguese services, "*the result of a natural meeting between the Asian commercial world and Western dynamism...*" [10].

In order to understand the cultural function of Macao during the Ming period, it is necessary to take both these basic social characteristics into account.

There were Asian networks of commercial interests and trading powers that were linked through partnerships and competition, which included the Chinese of 'the southern seas' (the *Nanyang*), Chinese from the provinces of Canton and Fujian, and the intermediary communities of Ryukyu and southern Japan/Kyushu. Groups of Portuguese maritime merchants, operating out of Malacca and Patane, began entering these Asian networks of inter-regional trade between Southeast Asia and the China Seas in 1512-1513, and between 1535 and 1555 Haojing/Macao became one of their favourite places.

The strength of these Asian networks of commercial interests and powers, which also now included a Portuguese element performing a crucial intermediary function in the external trade of China with Japan and Southeast Asia, led to Macao becoming an established port facility, playing a central role in the maritime trade network coordinated by the Portuguese.

---

[9] P. Adriano de las Cortes S.J. – *Viaje de la China*, c. 1626, ed. B. Monco, Madrid, Alianza, 1991, p. 97.

[10] L. Dermigny – *La Chine et L'Occident. Le Commerce à Canton au XVIIIème siècle, 1715-1833*, Paris Sevpen, vol.II, 1964, p.762.

Throughout the period from 1570-80 to 1630-40, Macao continued to consolidate its position in Ming-dynasty China at the heart of international maritime services in East Asia.

The maritime networks, which included partnerships, competition and conflicts, created the basis for the continued existence and economic viability of the port city of Macao, and made it a meeting point for people and merchandise coming from various parts of Asia and Europe.

II

The intercultural function of Macao during the Ming period was based on its situation as a port city and a common denominator for communities and interests that extended from the China Seas to the Atlantic Ocean.

It was this combination of networks and partnerships that made Macao one of the major centres for botanical and technological exchange in the 16th and 17th centuries.

Through the Portuguese and the Spanish, Ming-dynasty China soon adopted certain plants from the Americas, in particular groundnuts, sweet potatoes, maize and tobacco.

Macao and Manila were the major centres for the introduction of these crops, but in the case of groundnuts, it would appear that they were being planted in Guangdong province at an earlier date, possibly even from the time of the first Portuguese-Chinese contacts around 1516. In the 16th and 17th centuries they became an exotic hors d'oeuvre.

During the Ming period, China started growing sweet potatoes and maize, particularly in the provinces of Guangdong and Fujian. Tobacco also began to be grown successfully from around 1590-1600 in the provinces of Guangdong, Fujian and Zhejiang [11].

Throughout the Ming and Ch'ing dynasties, Macao remained one of the major centres for the introduction into China of a wide variety of other plants such as the potato, cassava, red pepper, papaya, tomato, guava, and cocoa, as well as green vegetables like beans, cabbage, lettuce and

---

[11] See, among other studies, K.C. Chang (ed.) – *Food in Chinese Culture: Anthropological and Historical Perspectives*, New Haven, Yale U. Press, 1977; D. Johnson, A.J. Natham, E.S. Rawski (ed.) – *Popular Culture in the Late Imperial Age*, Berkeley, U. California Press, 1985; E.N. Anderson – *The Food of China*, New Haven, Yale U. Press, 1988; and S.A.M. Adshead – *China in World History*, London, Macmillan, 1995, as well the bibliography therein concerning these introductions.

watercress, the latter being called 'the vegetable from the Western seas' in Cantonese to this day.

Around 1620-1630, in particular with the establishment of Manuel Tavares Bocarro's foundry[12], Macao became a major centre in Asia for the production of cannons, with Ming-dynasty China using either *fo-lang-chi* ("Portuguese-style" cannons, known at the time as *falcões* in Portuguese), or *Kung chung* ("Western-style" cannons):

> "... *in the said city of Macao there are seventy-three iron artillery pieces... since in this place there is one of the best foundries in the world, for bronze, as formerly used, as well as for iron...*" [13].

In addition, Western medicine and its instruments, printing using Western-style movable type characters, European publications, and the mechanical clock, were all introduced into China, as well as Japan, via Macao. In 1583, an Indian clockmaker, who had learned his art from the Portuguese, travelled from Macao to Zhaoqing in order to make an "iron clock" for the Prefect Wang Pahn [14].

The importance of Macao as an intercultural frontier between China and Europe, involving the coasts of East Asia and of the Atlantic and Pacific of the Americas, can also be seen on the linguistic level.

Every city functioning as a port of call for international trade develops a strong multilingual character. During the Ming period, Macao was the focal point for Chinese of different origins and cultural backgrounds and for other Asians such as Japanese, Malays, Vietnamese, the people of Siam, Indians and Koreans. Through merchants and missionaries, mainly Jesuits, it also became an international European city; Latin peoples in particular, Portuguese, Italians, and Spanish, but also English and Dutch among others, passed through or resided there.

---

[12] See N. Valdez dos Santos – *Manuel Bocarro o Grande Fundidor* in Bulletin no. 3 of the Museu e Centro de Estudos Marítimos de Macau, Macao, 1990, pp. 15-102 and the bibliography, in particular C.R. Boxer.

[13] António Bocarro – *O Livro das Plantas de Todas as Fortalezas, Cidades e Povoações do Estado da Índia Oriental* (1635), ed. Isabel Cid, Lisbon, I. Nacional, 1992, vol.II, p.264.

[14] Matteo Ricci S.J. – *Storia dell' Introduzione dello Cristianesimo in Cina*, ed. Pasquale M. Delia, *Fonti Ricciane*, Roma, Libraria dello Stato, vol.I, 1942, pp.201-212. On the history of the clock in Ming-dynasty China and the role of Macao, see: Carlo M. Coppolla – *Tecnica, Societa e Cultura*, Bologna, Il Mulino, 1989 and David S. Landes – *L'Heure qui est: les Horloges, la mesure du temps et la formation du Monde Moderne*, Paris, Gallimard, 1987.

The work that best expresses this multilingual element in Ming-dynasty China is the ***Dicionário Português-Chinês*** (Portuguese-Chinese Dictionary). This is a collective manuscript work, of which the only known copy belonged to M. Ruggieri (1543-1607), who brought it to Rome from Macao in 1588 [15].

Before and after the ***Dicionário Português-Chinês***, with no title, author or date, there was a collection of manuscripts, consisting of material of a diverse nature, mainly in Chinese characters. The only element that indirectly dates part of this collection refers to folios 12-16v, which were written in Zhaoqing in 1586. This section contains:

> "... *a text in Chinese characters that is a brief exposition of the fundamental beliefs of the Catholic faith and of the sacrament of baptism written by a missionary to a Chinese man of letters...*" [16].

Based solely on this passage, which dates the text found between folios 12-16v only, Paul Fu-Mien Yang has extended the place and date to cover folios 32 to 156, putting forward the theory that the ***Dicionário Português-Chinês*** was "probably" compiled in Zhaoqing between 1583 and 1588.

Our theory as to the date and in particular the place of production of the ***Dicionário Português-Chinês*** is based on a semantic analysis of the text. What we have is a Portuguese-Chinese vocabulary, with a central column of phonetic transcriptions, based on subjects such as nautical terms, the mercantile world, political and diplomatic relations, everyday life and certain aspects of Chinese culture and society.

Such subjects, especially those related to shipping, trade and finance, point to Macao as the origin of this dictionary. Also, many of the approximately three thousand words and phrases in Portuguese are specialised terms with connotations associated with aspects of life in Macao, such as "pão d'ouro" (literally 'gold bread', used to denote minimum basic food, 'daily bread') in fol. 126v, "*espingarda*" (musket) in fol.79, "*bombarda*" (short cannon), and "*ouvidor*" (type of magistrate) in fol.125v, an office that was created in Macao in 1578. These indicate a varied but particular realm of the Portuguese language, only accessible to an overseas community of Portuguese, that is Macao.

---

[15] Jap. Sin. I, 198, fols. 32a to 156 can be found in the Archivum Romanum Societatis Iesus in Rome.

[16] Paul Fu-Mien Yang S.J. – *The Portuguese-Chinese Dictionary of Michelle Ruggieri and Matteo Ricci: A Historical and Linguistic Introduction*, in Proceedings of the Second International Conference of Sinology, Academia Sínica, Taipei, 1989, p.204.

Analysis of the folios of the Portuguese-Chinese Dictionary would appear to indicate Macao and possibly Zhaoqing as the centres of production, the basis of the dictionary, both in Portuguese and Chinese terms, being associated with the maritime, mercantile, political and cultural life of Macao.

As to the question of dates, we know that Alexandro Valignano sought to study Chinese in Macao in 1577 and that M. Ruggieri began his studies there between 1579 and 1583.

The ***Dicionário Português-Chinês*** now in the archives of the Society of Jesus in Rome was written by various Western and Chinese hands. One theory is that work on the dictionary began around 1580 in Macao and continued, until 1588 at the latest, in Zhaoqing and Macao. It may also have been the linguistic guide to China that, compiled in Macao in 1582 at the latest, M. Ruggieri took to Zhaoqing in 1583, possibly continuing work on the phonetic transcriptions there.

The ***Dicionário Português-Chinês*** is essentially a product of Macao dating from the beginning of the 1580s, with additional work possibly being produced in Zhaoqing during the years 1583 to 1588.

The question of time and place is also linked to that of authorship.

> *"... The Dicionário Português-Chinês is a collective work, produced in Macao by groups of Portuguese representing commercial, political and religious powers. A vocabulary that was accumulated thanks to the innumerable contributions from educated Chinese, Portuguese and Chinese merchants and seamen, and Jesuit missionaries... missionaries who organised this information into a didactic tool..."* [17].

This collection of over two thousand terms, with no title, author or date, is the first Chinese dictionary for a Western language. The entries in Portuguese, both words and phrases, range from "*aba da vestidura*" (border of garment) in fol.23 to "*zunir a orelha*" (to buzz in the ear) in fol.156, and are followed by two columns, the first giving the phonetic transcription and the second the term in Chinese.

While it is a vocabulary ranging from A to Z, it is far from complete, being essentially a practical and informative collection in which nautical terms dominate, closely followed by words from everyday life and the commercial and political worlds.

---

[17] Luís Filipe Barreto – *A Dinâmica Cultural de Macau c.1560-c.1660, I Parte – A Fronteira Cultural* – in Macau, 2nd series, no. 58, February 1997, pp.52-54.

The following words and phrases are given merely as examples of the preponderance of vocabulary related to shipping, which lends weight to the idea of the port of Macao as its source: *almadia* (type of long narrow boat) in fol.40, *a nao* (ship), ancora (anchor) *ancorar* (to anchor) in fol.42, *a remos de vela* (with the oars of a sailing ship), *arimar o outro* (to come alongside another ship) in fol.45v, *barcada* (boatload), *barra de naos* (harbour entrance) in fol.50v, *batel* (skiff) in fol.52, *borda* and *borda de nao* (gunwale of a ship) in fol.53, *cabo* (cape) and *cabo do mundo* (the end of the world) in fol.55, *caraca* (carrack) in fol.57, *carta de marear* (navigation map) in fol.57v, *cobertura* (canopy) in fol.58v, *tormenta* (storm) in fol.150, *vento à popa* (before the wind) in fol.135v, and *verga de nao* (ship's yard) in fol.154.

A further sample comes from the language of commerce, once again associated with the life of the city of Macao: *açucar rosado* (raw sugar), *açucare candil* (crystallized sugar), *açucar refinado* (refined sugar) in fol.35, *afina ouro* (to refine gold), *afinar prata* (to refine silver) in fol.36v, *barra de ouro ou de prata* (gold or silver ingot) in fol.50v, *canela* (cinnamon), *canfora* (camphor) in fol.56v, *cravo* (clove) in fol.36v, *prata* (silver), *prata fina* (pure silver) in fol.133v, *seda* (silk) in fol.143, *sandalo* (sandalwood) in fol.142v, *aforrar* (to save money) in fol.36v, *cambar permuta* (exchange and barter) in fol.56, *dar credito* and *dar fiado* (to give credit) in fol.72v, *mercar* (to trade), *mercador* (merchant), *mercadoria* (goods), and *mercado* (market) in fol.118v.

The ***Dicionário Português-Chinês*** resulted from the combined knowledge of Portuguese and Chinese merchants and Jesuit missionaries, particularly M. Ruggieri, as well as educated Chinese and translators such as the Chinese Christian Filipe Mendes, who was from, or at least had close links with, Macao. There may also have been young people of Chinese or Portuguese-Chinese origin who had links with merchants or Jesuits, as well as lay brothers, such as the Macao-born Sebastião Fernandes/Zhong Mingren (1561/2-1622), who was one of M. Ricci's Chinese teachers and the principal author of the second Portuguese-Chinese Dictionary in 1598 [18]. He began his novitiate in the Society of Jesus in 1591, but may already in the 1580s have been involved with Ruggieri and Ricci in their informal

[18] "... I piu vecchi della missione, et avere selo il fratello Bastiano che sapeva molto bene la lingua della cina, fecerro un bello vocabulario, e messero in regola et ordine le cosa di questa lingua; con che da li avanti il doppio piu facilmente si poteva imparare..." Matteo Ricci S.J. *Storia dell'Introduzione dell Cristianesimo in Cina*, ed. Pasquale M. Delia, *Fonti Ricciane*, Roma, Libraria dello Stato, vol.II, 1949, p.32.

studies of the Chinese language, since in 1589, together with Francisco Martins, he was one of the first Chinese applicants to the Society [19].

The ***Dicionário Português-Chinês*** also served another function, one which has always been a factor of life in Macao, through its phonetic transcriptions of Chinese characters. The transcriptions vary between those from Mandarin Chinese, either the northern dialect from around Beijing or the southern form from around Nanjing, or from Cantonese or Hokkien, the dialect spoken in Fujian.

Building no doubt on an already existing legacy, Macao increasingly became a magnet and meeting point for Chinese from different regions and communities, social strata and cultural backgrounds.

It should be noted that the variety shown in the phonetic transcription of Chinese characters, the result of contact with different Chinese communities, is yet another indication that Macao rather than Zhaoqing was where the dictionary was produced, as well as providing further evidence that it was a collective work.

The dual name Haojing/Macao shows that the Portuguese were involved in partnerships with Chinese circles from Canton and Fujian who had already established relations with the Kingdom of Ryukyu, southern Japan and Southeast Asia before the arrival of the Portuguese; this was in fact one of the main reasons for establishing and maintaining a Portuguese maritime-mercantile community.

For example, in the codex Jap.Sin. I 198, folio 169, in the pages following the dictionary, there is the term "Maquao" followed by the Chinese character for Haojing/"Oyster Mirror", with no phonetic transcription. By 1588 at the latest, Haojing/Maquao are equivalent terms corresponding to Mandarin and the phonetic transcription from Hokkien respectively, which illustrate the main partnerships of the Portuguese merchants.

---

[19] In 1593, A. Valignano states that Sebastião Fernandes is "... Chinese though born and raised in Macao and has Portuguese blood...". The "*rol das casas*" for 1592 describes Sebastião Fernandes and Francisco Martins as follows: "... they are both Chinese, though born and raised among Portuguese", in Monumenta Historica Japoniae I, ed. J. Franz Schutte S.J., Rome, M.H. Iesu, 1975, pp. 286 and 325. See Frederico Masini – *Some Preliminary Remarks on the Study of Chinese Lexicographic Material Prepared by the Jesuit Missionaries in China in the Seventeenth Century* in *Western Humanistic Culture Presented to China by Jesuit Missionaries (XVII-XVIII centuries)*, ed. F. Masini, Rome, I.H.S.H., 1996, pp.235-245.

The ***Dicionário Português-Chinês*** of the 1580s was also the forum *par excellence* in Macao for the meeting of lay and clerical cultures, of the practical and the theoretical. It was a collective effort arising from the meeting of interpreters, merchants and missionaries, and of Europeans of Latin origin with Chinese, at least one of whom was from Macao.

III

The cultural landscape of Macao during the Ming period is divided into two major areas: the lay culture, essentially technical and practical, and the clerical culture, more theoretical and erudite. Both were the result of maritime-mercantile dynamism and, more directly or more indirectly, respectively, served the basic interests and cultural life of the port. Thus, the two areas found themselves interlinked in certain strategic areas of involvement and common interest, such as the training of interpreters, the systematisation and the practice of translation, and the process of botanical and technological transfer.

In the Ming period, all lay culture was in manuscript form and the known texts from the west side of Haojing are written in Portuguese.

Lay texts were written in Portuguese, although authors sometimes signed their work with Chinese characters since they did not know the Portuguese equivalent:

> *"Here is my signature. It is in Chinese since I do not know the Portuguese..."* [20].

This can be seen in the example of the Chinese from Macao whom we know only by the Portuguese name of Salvador Dias; in 1626 he wrote two items of strategic maritime information about Dutch competition and the island of Formosa:

> *"... concerning the open trade between the Chinese and the Dutch who, unless they remedy the situation in time, will no longer have it..."* [21].

These reports from Salvador Dias represent summaries written in Portuguese of information that was originally collected and written in Chinese.

---

[20] Information from Salvador Dias in *Documentação Ultramarina Portuguesa*, Lisbon, C.E.H.U., 1962, vol.II, p.68.
[21] Idem, p.65.

A constant factor in the lay culture of Macao was the collection, organisation, processing and translation of material and information from Chinese texts, although there were also some from Japanese and Annamese, among other languages. These texts and data, especially of a nautical, cartographic, or geographical nature, were appropriated and utilised.

Portuguese knowledge of the seas and coastlines of China produced in Macao is based to a large extent on this ability to collect, select, translate and incorporate Asian knowledge. Macao was an intercultural frontier post through which Asian knowledge was transferred into the body of European knowledge of East Asia.

The majority of these manuscript texts written in Portuguese give no indication of authorship since they were collective works, used by all and subject to constant addition, amendment and correction on the part of specialised sectors of the maritime-mercantile community of Macao.

Today, faced with a map, a navigational guide, an account of a voyage, or a geo-political commercial report which bears no signature, the historian classifies it as 'anonymous', but this is generally an anachronistic simplification that may obscure the true situation.

In the 16th and 17th centuries, the notion of intellectual property was only just beginning to emerge as the dynamic of capitalism became increasingly important. Maps, ships, guides, reports and information were all seen as goods in a world in which the roles played by corporations, the family, and the community were fundamental. These goods were constantly being collected, added to and amended, and there was little place for the idea or the reality of the individual, or of know-how as private property.

This whole practical culture of a nautical and mercantile nature was a collective culture, created by the community of merchants and seamen who used and improved upon the 'goods' in their everyday professional lives in maritime trade.

It was a pragmatic and empirical culture that developed in close relation with shipping and trade, with these cultural 'goods' being produced solely by and for these circles. It is this intimate link between maritime trade and technical and practical knowledge of seamanship that led to the existence in Macao of two major elements in the lay culture.

The first, from around 1560-1570 until the beginning of the 17th century, consists mainly of navigational guides from Macao to Japan and of guides, maps and other information relating to the coastal areas of Guangdong and Fujian provinces:

"... *note whoever leaves Macao for Japan in this June monsoon that, if the weather is good, they cannot turn windward of the rock that looks like a sail...*" [22].

The second element, beginning around 1565-1575 and which by 1640-1650 had become clearly the dominant one, was mainly devoted to navigational guides from Macao to Southeast Asia, creating

"... *a new generation of guides for the coasts of the Indochinese peninsula, a result of the continual growth in trade between Malacca and Macao... the routes between Macao and the Gulf of Tonkin through the Hainan Strait were the subject of the first known guides to this region...*" [23].

These efforts and cultural emphases accompanied changes in maritime trade and the political situation in East Asia and Ming-dynasty China. From 1639-1655, the so-called 'southern ports' (Makasar, Timor, Solor, Flores) replaced Japan as destinations out of Macao, coinciding with the fall of Malacca in 1641 and of the Ming Dynasty in 1644:

"... *leaving on muleback to Malacca, you will leave from the western quay and once at sea you will pass by the islands...*" [24].

The lay culture of Macao during the Ming period can be grouped into four areas:

1 – nautical guides and descriptions;
2 – cartography;
3 – reports and strategic information of a commercial or political nature;
4 – information on depths and currents, together with general reports, particularly on the coastal areas of Guangdong and Fujian provinces.

Thus the lay culture of Macao is dominated by seafaring, with navigational guides, maps and coastal surveys making up over 85% of production. This emphasis on knowledge of the seas in the lay culture of Macao is evidence of the fundamental importance of the link to maritime trade for the life of the city:

---

[22] *Roteiro de Macao pera Japao* (c.1570-1580), Bibliothèque Nationale, Paris, cod. port., no.58, fol.77v, in L. Bourdon and L. Albuquerque - *Le "Livro de Marinharia" de Gaspar Moreira*, Lisbon, J.I.U, 1977, p.122.

[23] Pierre-Yves Manguin – *Les Portugais sur les Côtes du Viêt-Nam et du Campã*, Paris, E.F.E.O., 1972, pp.61-62.

[24] *Roteiro de Macau para Malacca na Monção de Janeiro e Fevereiro e em todo o tempo* (c.1570), Casa Cadaval, MS. 972, fol.78v in Pierre-Yves Manguin – *Les Portugais sur les Côtes du Viêt-Nam et du Campã*, Paris, E.F.E.O., 1972, p.258.

> *"... the wealth of Macao lies in the sea, and the whole city lives from it, there are no permanent goods other than those that the winds and the seas bring; without these, there is nothing..."* [25].

This manuscript culture, written in Portuguese, arose mainly from the translation and incorporation of nautical knowledge of the China Seas, as can be seen, for example, in the anonymous ***Carta de Macau e do Delta de Cantão*** (Letter from Macao and the Canton Delta) dating from around 1643 [26]. Written in Macao and sent to Lisbon, probably in 1646, it is a typical letter from China with great attention paid to the interior, with the coastline and the sea as the frontier. Within the area between Hainan and Lantao, the letter focuses on the region between Canton and Macao. Place names and explanatory notes of a commercial or strategic nature abound. For example, in Canton, adjacent to the "*pagode do meyo do Ryo*" (pagoda in the middle of the river), there is a note:

> *"... on this river there are usually two or three thousand ships"*.

The practical and technical culture of seafaring was a constant feature of this intercultural frontier, through the collection, translation and incorporation of Asian nautical guides and cartography into the body of knowledge of Europeans based in Asia:

> *"... leaving the port of Macao, beyond the islands, three or four leagues out to sea, head south-southwest until you have sight of the coast of Chanpa [South Vietnam]..."* [27].

IV

It is, however, the clerical erudite culture, which in Macao during the Ming period was dominated by the Jesuits, that best reflects the intercultural dimension of the port.

---

[25] Luis da Gama S.J., *Carta de Macau* dated 15 December 1664, B. Ajuda, Jesuítas na Ásia, cod.49-IV-56, fol.379.

[26] See copy of the map from B. da Ajuda in Luís Filipe Barreto – *Cartografia de Macau, séculos XVI e XVII*, Lisbon, Missão de Macau, 1997.

[27] *Roteiro das Ilhas de Cantão pera Malaqua*, c.1570-1580, B.N. Paris, cod. port., no.58, fol.77 in L. Bourdon and L. Albuquerque – *Le "Livro de Marinharia" de Gaspar Moreira*, Lisbon, J.I.U, 1977, p.121.

The first translations from Latin into Mandarin and from Mandarin into Latin were produced in Macao, even if at a later date the work may have continued in Zhaoqing.

At the end of 1581 or beginning of 1582, the first catechism in Chinese was produced in Macao. It was a collective work by the Jesuits M. Ruggieri and Pêro Gomes, together with an unnamed educated Chinese:

> *"... we arrived on the eve of São Tiago (St. James) of 81... where we encountered another priest, Miguel Rogério, from the China mission... he produced a catechism that an educated Chinese Christian translated into Chinese here in Macao... the priest had the catechism printed in Chinese characters, which was then distributed throughout China..."* [28].

According to Francisco Pires, it would appear that the ***Catecismo/Tian Zhu Shi Lu*** was ready and printed in 1581, but a letter from one of the authors, Pero Gomes, written in Macao on 25 October 1581 refers to the work as yet to be printed, or even translated:

> *"Father Rugerio and I, during the months I have been here, have been working on a brief history of the beginning of the world, which will also serve as Christian doctrine, in the form of a dialogue, to be translated into the Chinese language..."* [29].

In November or December 1581 or at the beginning of 1582, a manuscript copy of the ***Tian Zhu Shi Lu*** existed in Macao which, according to the Jesuit Pêro Gomes, was printed and immediately sent for distribution in China.

The work is a short text concerning the existence and the nature of God, the immortality of the soul, natural law, and the sacrament of baptism. It was an exposition of Christianity through Buddhism and probably the first printed work in Chinese of Christian literature, since there are no known Chinese texts from early medieval Christianity.

The only copies of the ***Tien Zhu Shi Lu*** known today are of an edition printed in Zhaoqing in 1584 [30]. This had led some to believe that the ***Tien***

---

[28] Francisco Pires S.J. – *Pontos do que me alembrar*, Macao, 1622-1623, in Monumenta Historica Japoniae I, ed. J. Franz Schutte S.J., Rome, M.H.S.J., 1975, p.386.

[29] Pêro Gomes S.J. – *Carta de Macau para Roma*, 25 October 1581, in Monumenta Historica Japoniae I, ed. J. Franz Schutte S.J., Rome, M.H.S.J., 1975, p.117.

[30] See J. Gernet – *Sur les Differentes Versions du Premier Catechisme en Chinois de 1584*, Studia Sino Mongolica, 1979, pp.407-416, and Pascale Girard – *Os Religiosos Ocidentais na China na Época Moderna*, Macao, F.M., 1999, pp.278-279.

Zhu Shi Lu was produced in manuscript form in Macao and only printed from wood blocks in Zhaoqing in 1584. However, there is no evidence or reason to think that a work concluded at the end of 1581 or the beginning of 1582 should have remained in manuscript form; moreover, the document written by Francisco Pires gives us to understand that it was printed in Macao.

Whether in manuscript or printed form, the fact is that the first Chinese translation of a European work was produced in Macao by 1582 at the latest.

The first exercises in translating Chinese into Latin must also date from around this time, carried out by a team coordinated by M. Ruggieri.

M. Ruggieri had been in Macao since July 1579 learning to read, write and speak Chinese. In 1582, he translated and interpreted for A. Valignano certain passages from Chinese books acquired by the Jesuits, particularly "*... Chinese geographical and historical treatises and a dictionary of plants...*" [31].

First in Macao and later in Zhaoqing, between 1579 and 1588, M. Ruggieri translated the first part of the ***Da Xué/The Great Teaching*** as part of his studies of the Chinese language.

In Rome in 1593, Antonio Possevino published certain parts of M. Ruggieri's translation of Confucius, the idea being to produce a ***Liber Sinensium*** of moral principles [32].

The first translation from Chinese into Latin produced by the Jesuits' team of translators was the ***Da Xué/The Great Teaching***. The ***Da Xué***, together with ***Chu Hsi*** (1130-1200) and the neo-Confucianist movement, represented the basic exposition of Confucianism and became the first text to be used in the curriculum of the Imperial School and the central book in Civil Service examinations.

Macao was the frontier where two major languages and cultures met, one European, the other Oriental. The translations between Latin and Mandarin demonstrate the efforts made to forge links between Christianity and Chinese Buddhism and Confucianism. They also served the purposes of the missionaries in their desire to know more about, and be better known by, Ming-dynasty China.

---

[31] A. Valignano S.J. – *Sumario de las cosas de Japon* (1583) and *Adiciones del Sumario de Japon* (1592), ed. J.L. Alvarez – Taladriz, Tokyo, Sophia University, 1954, p.175.

[32] See Knud Lundbaek – *The First European Translations of Chinese Historical and Philosophical Works*, in *China and Europe: Images and Influences in the Sixteenth to Eighteenth Centuries*, ed. Thomas H.C. Lee, Hong Kong, Chinese U. Press, 1991, pp. 29-43.

The intercultural role of erudite culture in Macao was not limited to such translations, nor to the establishment of the first libraries of European books in China and of Chinese and Japanese books by Europeans. It can also be seen in the links established between Chinese, Japanese and European cultures as found in the printed works of the Society of Jesus, which had an impact in both Asia and Europe [33].

In July 1588, a printing press using Western-style movable type arrived in Macao on its way from Portugal to Japan. In the same year, ***Christiani Pueri Institutio***, by the Jesuit priest Giovanni Bonifacio, was published, having originally been printed in Salamanca in 1575. It is a short Latin treatise to teach young people Christian doctrine, which appears to have been used as a textbook to teach Latin and to spread Christian ethics, especially in Japan, as can be assumed from the letter "to the students in Japanese seminaries", written by A. Valignano, which appears at the beginning of the work.

The work ***De Missione Legatorum Iaponensium ad Romanorum Curiam... Dialogus***, whose principal authors were the Jesuit Duarte de Sande (1547-1599) and A. Valignano (1539-1606), was published in 1590, "in the port of Macao in the Kingdom of China" (in the words of the publication itself).

The ***De Missione*** is in the form of a dialogue, designed to introduce Europe and Christianity to Japan. It emphasised all the positive aspects, but particularly nautical matters, mechanical clocks, printing and the university, "*... a notion of things European and of Christianity, which at the time were little known in Japan...*" [34].

However, in the thirty-third colloquy, the ***De Missione*** also gives an introduction to Ming-dynasty China based on "*... the reality described by the Jesuit priests living in China...*" [35].

The printing press returned from Japan to Macao in 1614, and in 1620, it was used to print the ***Arte Breve de Lingoa Iapoa*** (Short Grammar of the Japanese Language) by the Jesuit João Rodrigues Tçuzzu (1561-1633). This is partly a shorter version ("*... to produce this short extract from the longer grammar, which will give beginners an introduction to the*

---

[33] See J. Maria Braga – *The Beginnings of Printing at Macao*, Lisbon, Studia, 1963, no.12, pp.29-137.

[34] Duarte de Sande S.J. – *Diálogo sobre a Missão dos Embaixadores Japoneses à Curia Romana* (1590), ed. A. Costa Ramalho, Macao, F. Oriente, 1997, p.20.

[35] *Ibid.*, chap.XXXIII, p.329.

language")[36] of the work ***Arte da Lingoa de Iapam*** (Grammar of the Japanese Language), Nagasaki 1604-1609, coordinated by the same author. But it is also "... *a much better work as a grammar and in many respects is completely new... it is probable that Rodrigues had more free time in Macao to work on a better plan and to organise his ideas more clearly...*" [37].

The printed works of the Jesuits in Macao, in Latin and Portuguese, were mainly destined for the Japanese elites with links to Christianity, but they also circulated in the rest of Asia, in America and in Europe.

To illustrate the circulation network of works printed in Macao, we have chosen just two examples. A translation from the Latin into English of the Ming-dynasty China section of ***De Missione Legatorum*** by the Jesuit Duarte de Sande appeared in London in 1599, in the second volume of the second edition of R. Hakluyt's ***Principal Navigations***, with the title ***Excellent Treatise of the Kingdom of China, and of the Estate and Government... Printed in latine at Macao, a citie of the Portugals in China*** [38].

The translation into English of this part of the work printed in Macao in 1590 played an important role in adding to the body of information on China in English. At the time, all such information was the result of knowledge acquired and transmitted by the Portuguese, Spanish and Italians to the rest of Europe, a process which had begun in 1577 with the English publication of Galeote Pereira [39].

Thus, at the end of the century, in 1599, the fourth book on China in English was published, taken from the writings of the Jesuit Duarte de Sande, who had lived in Macao for five years, with a period in Zhaoqing from August 1585 to November 1587, and who had excellent access to abundant, up-to-date and accurate information on Ming-dynasty China.

The second example of the impact in Europe of works printed in Macao concerns the ***Arte Breve da Lingua Iapoa***. The 1620 Macao edition served as a guide for the third European Japanese grammar, the ***Ars Grammaticae Iaponae Linguae*** (Rome, 1632) by the Dominican Diego Collado.

---

[36] *Arte Breve da Lingoa Iapoa tirada da Arte Grande da mesma Lingoa*, Joam Rodriguez da Companhia de Jesu, Macao, 1620, Introduction.

[37] Michael Cooper S.J. – *Rodrigues o Intérprete: Um Jesuita no Japão e na China*, Lisbon, Quetzal, 1994, p.239.

[38] See *Um Tratado sobre o Reino da China*, ed. Rui Manuel Loureiro, Macao, ICM, 1992.

[39] Among many other studies, see Donald F. Lach – *Asia in the Making of Europe*, vol.I, book II, Chicago U. Press, 1965, pp.730-821.

Works printed in Macao during the Ming period circulated in Europe, and in London and Rome they made a major contribution to the growth of Western knowledge about China and Japan. The same can be said of manuscript works, from the vast production and circulation of letters to the manuscript navigational guides to the China Seas produced in Macao, some of which were printed in Amsterdam in 1595 by J.H. van Linschoten – ***Reysgheschrift van de Navigatien der Portugaloysers***. Through manuscript copies, translations and printed works, Macao was a centre for mutual exchanges between Europe and East Asia during the Ming period.

The erudite culture of Macao was not limited to printed works since manuscript documents were far more numerous. Many were books that were finished, revised, and compiled in Macao but which in general had been begun in other parts of Asia, from Japan and China to India.

This means that the resources existed in Macao, in terms of time, information and libraries, that enabled these works to be concluded. We will give just four examples of this.

In Macao in 1582, between 9 March and 31 December, A. Valignano revised the ***Sumario de las Cosas... de la Yndia Oriental*** (Summary of Things... of Eastern India)[40], written in 1579, and began writing the ***Sumário de las Cosas de Japon*** (Summary of Things of Japan)[41], which he completed in Cochin in 1583. Between 1592 and 1594, the Jesuit Luis Frois completed his ***História de Japam*** (History of Japan)[42] in Macao, a large part of which had been written in Japan between 1585 and 1591. In Macao, from 1620 to 1627, the Jesuits João Rodrigues Tçuzzu and Martinho Mara wrote their ***História da Igreja do Japão*** (History of the Church of Japan)[43] in Macao, based on documents brought there in 1619.

The clerical erudite culture of Macao during the Ming period can be grouped into four types. Between thirty and forty per cent of these works are

---

[40] A. Valignano – *Sumario de las Cosas que pertencen a la Provincia de la Yndia Oriental* (1579) in *Documentação para a História das Missões do Padroado Português do Oriente*, ed. A. Silva Rego, Lisbon, A.G.U., 1958, 12th vol., pp.470-638. See also A. Valignano S.J. – *História del Principio y Progresso de la Compania de Jesú en las Indias Orientales (1542-64)*, ed. J. Wicki, S.J., Rome, B.I.H.S.I., 1944.

[41] A. Valignano S.J. – *Sumario de las Cosas de Japon* (1583), ed. J.L. Alvarez-Taladriz, Tokyo, Sophia University, 1954.

[42] Luis Frois S.J. – *Historia de Japam* ed. J. Wicki, S.J, Lisbon, B. Nacional, 1976-1984, 5 vols.

[43] João Rodrigues Tçuzzu S.J. – *História da Igreja do Japão*, ed. Abranches Pinto, Macao, Notícias de Macau, 1954.

geographical or anthropological treatises on Asia, particularly Japan. Slightly over twenty per cent are reports and accounts of missionary experiences in Japan and China, such as the 1614 work of the Jesuit Afonso de Lucena ***De Algumas Coisas que pertencem à cristandade de Omura*** (Of some things relating to Christianity in Omura)[44] or the above-mentioned (in note 28) "***Pontos do que me alembrar***" (Points I remember) of the Jesuit Francisco Pires, dating from around 1622-23.

A further twenty per cent are texts relating to the internal organisation of Franciscan, Augustine and Jesuit missions, in the form of '*catálogos*' (catalogues), '*rol de casas*' (mission registers), reports and other information. The remaining thirty to forty per cent are polemical or doctrinal works in dialogue form directed at the Asian world, and different European views on Asia. By way of example, we would simply cite the ***Apologia de la Compania de Jesus de Japon y China*** (Apologia of the Society of Jesus of Japan and China), Macao, 1598, by Alexandro Valignano[45] and the report sent by Martim Inácio de Loiola to Philip II of Spain from Macao in 1587 [46].

In contrast to the technical and practical culture, erudite culture in Ming-dynasty Macao appeared in various languages (Portuguese, Latin, Chinese and Spanish) in both manuscript and printed form. The variety of languages, or at least Western ones, arose from the fact that the missionaries, and in particular the Jesuits, were an international elite open to other nationalities, both Eastern and Western.

There are many other aspects of the cultural landscape of Macao during the Ming period that could be studied, particularly the epistolary network that linked Macao to China and Japan, as well as to Manila, Malacca and Goa, and through them to cities in America and Europe – Lisbon, Madrid, Rome, London and Amsterdam, among others.

However, the body of works described above is already sufficient to appreciate the intercultural dimension of Macao during the Ming period. From the establishment on Chinese territory of the first community of Portuguese merchants in the years 1555-1557, followed by a Jesuit mission

---

[44] Afonso de Lucena S.J. – *De algumas cousas que ainda se alembra o P. Afonso de Lucena que pertencem a Christandade de Omura (1578-1614)*, ed. J. Franz Schutte S.J., Rome. I.H.S.J., 1973.

[45] A. Valignano S.J. – *Apologia de la Compania de Jesus de Japon y China (1598)*, ed. J.L. Alvarez-Taladriz, Osaka, 1998.

[46] In *Sinica Franciscana, Relationes et Epistolas Fratrum Minorum Saeculi XVI-XVIII*, ed. Van Den Wyngaert, G. Mensaert and F. Margiotti, Quaracchi – Rome, 1936-1975, 8 vols. (The text quoted can be found in II, pp.210-213.)

in 1565, the port gradually became the centre for Ming-dynasty China's international relations and for European contact with East Asia.

Through its multiple partnerships and unofficial shared governance, together with the practical-lay and erudite-clerical cultures, the city became the intercultural frontier *par excellence* between Europe and East Asia from the mid-1500s to the mid-1600s.

During the Ming period, Macao functioned as a bringer of news of China and Japan to Europe and as an introduction to Europe for China and Japan. At the time, it was an open border between Europe and China, through which not only goods, but also languages, books and ideas, were exchanged.

Translated from Portuguese by Paul Covill.

SURVEY AND STUDY OF PRE-1900 CHINESE MAPS SEEN IN EUROPE

LI Xiaocong
Department of History, Beijing University

Since the end of the fifteenth century, the newly emergent Capitalism of Western Europe began to grow rapidly, overcoming the shackles of feudalism, and pounding at age-old civilizations all over the world. Demands that could not be met within European society created the need for new voyages of discovery and colonial expansion overseas. The first Europeans to reach Asia by sea were the Portuguese and the Spanish. Along with the Chinese silks and pottery brought back from East to West, one of the items which showed the history, geography and culture of the ancient Chinese empire, were "the so-called second language of geography" maps. From this point onwards, Chinese maps began to be collected in the West.

I. Chinese maps and the West

I.1. Introduction of chinese maps to the West

How did the Chinese maps spread to Europe?

In 1571, the Spanish established a new fort, the port of Luzon (Manila) on the western coast of Luzon in the Philippines, and opened a sea route from Fujian Province of Ming China to Europe via the Pescadores Islands (Penghu 澎湖), Luzon, Sulu Archipeligo, and the Moluccas. Just 2 years after 1571, a universal map of the Ming Empire in the Chinese by woodcut, *Gu jin xing sheng zhi tu* (古今形胜之图), "A map of Sites and Topographies, Past and Present" which had been reprinted by Jinsha shuyuan in Longqixian, Fujian Province (福建龙溪县金沙书院) in 1555, was carried along this route and presented, in 1574, by Guido de Lavezaris, the second Spanish Governor-General in Luzon, to the Spanish King Philip II in Madrid. This map, plus letters, is still kept at the Archivo General de Indias,

Sevilla. Although it did not gain any attention and had little influence on European cartography at the time, it is possible that this is the earliest Chinese map preserved in Europe.

About 1583, the Jesuits Michele Ruggieri, Matteo Ricci and others came from Italy into Guangdong Province via Macao, which had been occupied by the Portuguese since 1557. Ricci brought with him world maps from Europe and sent a few maps of China that he had compiled after consulting Chinese maps and local gazeteers to Macao in 1585. These maps in Chinese were carried to the Vatican by Portuguese ships and are still preserved at the Biblioteca Apostolica Vaticana. While Ruggieri was in Canton, he had a very strong interest in the *Guang yu tu* (广舆图), an atlas of the vast domains by Luo Hongxian (罗洪先), and he translated all the Chinese place names on the maps into Latin. This atlas was brought to Rome when Ruggieri returned to Italy in 1590.Italian scholars have recently recovered this atlas in the national archives and reprinted it as "*Atlante della Cina*" (Istituto Poligrafico e Zecca dello Stato, Roma, in 1993) based on the Latin manuscripts of Ruggieri. Because Ruggieri never visited North China and probably only knew Cantonese, he wrote the place names in Latin according to the dialects of Guangdong.

From the end of the 16th Century, the Dutch were engaged in struggles for independence against the Spanish in order to gain more profit from trade. The Dutch attacked the Spanish colonies in Asia and America, and their fleet embarked on a virgin voyage to Java. In 1602, they established the United East India Company (de Verenigde Oostinsche Compagnie) at Batavia, composed of six Chambers of Commerce in Holland and Zeeland, to manage trade and colonial affairs. All of the reports and information of Dutch activities in Asia had to be gathered at the Headquarters in Batavia every year and were then handed over by Dutch ships to the Headquarters in Amsterdam to be checked and filed. Dutch activity in the 17th Century caused Chinese maps to spread to Holland first, rather than to Spain or Portugal, and then to other countries. For example, there is a copy of the first edition of *Guang yu tu* (广舆图 1555) in The Hague, copies of *Guang yu kao* (广舆考 1595) in Florence and St.Petersburg and a map of the Chinese Ming Empire entitled *Beizhi huang ming yi tong xing shi fen ye ren wu chu chu quan lan* (备志皇明一统形势 分野人物出处全览 1605) in Cracov. All were first brought to Amsterdam by Dutch ships and then moved to other places. Most of the Chinese maps collected in the Jesuit College at Clermont in Paris before 1764 came to Europe in the same way.

At the same time, however, Russia crossed the Ural Mountains to expand its territories and came into conflict with China. As a result, Chinese maps started to be carried to Russia and Europe by a route across Siberia. For example, when representatives from China and Russia held negotiations to delimit borders at Nepchu (尼布楚 Nerchinsk) in 1689, Jean Francois Gerbillon and Thomas Pereira, two Jesuit interpreters for the delegation of the Qing Empire, exchanged Chinese maps for fur and food with the Russians. Later these Chinese maps were carried back to Moscow and St. Petersburg.

The various Chinese maps carried to Europe became the source for the compilation of maps of Asia by European cartographers. Over a long period, the cartographers in Europe corrected their depiction of the geography of East Asia and started to draw the new World maps with more representative of the actual situation. Henceforth Jesuits and merchants returning from East Asia frequently brought Chinese maps as gifts to the Pope, Kings or friends.

In the 18th century, the French and the English successively supplanted the positions of Portugal, Spain and the Netherlands in East Asia and began to pursue trade and colonial expansion. Many military maps from the Chinese Qing Empire were plundered and removed to the West at this time. By the middle of the 19th century, almost all countries in Europe had had expeditions to China. A large number of Chinese maps was taken from China to Europe by missionaries, merchants, diplomatists, visitors and soldiers with the various missions. These Chinese maps preserved in Europe are rarely seen by the public.

I.2. Western studies on chinese maps

Western scholars started to introduce Chinese maps in academic journals in the second half of the 19th century. Since the beginning of this century, more and more scholars in China and abroad have paid attention to the value of these maps as cultural and scientific artifacts. One benefit of this research has been ascertaining where these maps are preserved.

The *Guan gyu tu* (广舆图) by Luo Hongxian (罗洪先) was the first Chinese map to be introduced by European scholars and had effected the European cartographers to know the East. Klaproth in 1814, Baldelli in 1827, Frescura and Mori in 1894, all mention a copy of 1595 that was carried to Florence by a local merchant, Francesco Carletti. They thought that the earliest edition of this map in Europe was a copy of the 1565 edition preserved in St.Petersburg, and this one was a copy of 1595. However, as far as I am aware, the map in Florence is not the *Guang yu tu*, but a derivative edition, the *Guang yu kao* (广舆考), an examination on the basis of the atlas by Wang Zuozhou. I believe the earliest edition of *Guang yu tu* in Europe to be the one that is kept in the Meermanno-Westreenianum Museum in The Hague.

In 1903 K. Ahlenius announced the existence of a copy of Verbiest's world map: *Kun yu quan tu* (坤舆全图 1674) held in the library of Uppsala University in Sweden, and in 1923 K. de Jaegher made a further description of it in the journal T'oung Pao.

In 1919, in an article entitled “Another Jesuit world map made in China”, J.F. Baddeley, introduced a map of the stars compiled by Schall von Bell (汤若望: *Huang dao nan bei liang zong xing tu* 黄道南北两总星图 ca.1630) which aroused special interest among scholars from Berlin, Paris and Rome.

The whereabouts of the Chinese world map *Kun yu wan guo quan tu* (坤舆万国全图) by Father Matteo Ricci are unclear: J.F. Baddeley (1917), E. Heawood (1917), L. Giles (1918), William Hong (1935) and D'Elia (1938) mention several copies from different periods preserved separately in the Biblioteca Vaticana, Biblioteca Milano, Royal Geographical Society in London and the Library of Kyoto University.

From 1933 to 1943, Walt Fuchs published several articles on the complete atlas of the Kangxi Era, the *Kang xi huang yu quan lan tu* (康熙皇舆全览图) compiled by the Jesuits and Chinese cartographers, as well as on other general maps from the Ming and Qing dynasties. Fuchs not only described copies held in China, but also those preserved in Tokyo, Kyoto, Seoul, Paris and Berlin. He showed the styles and sources used in general maps before the middle of Qing Dynasty by means of photographs.

After the 1950's, Western scholars maintained their interest in the Chinese maps produced by the Jesuits, and also began to research other Chinese maps taken overseas, revealing more information about old maps held outside China.

In 1954, J.V. Mills depicted the historical sources and mapping features of the coastal maps in the Ming and Qing Dynasties, in his essay on Chinese Coastal Maps in Imago Mundi (No.11). He also studied and examined 12 manuscripts and printed maps which are included in the Library of Congress, Washington D.C., British Library, Royal Geographical Society and Royal Asiatic Society in London, and the Collection of L. Bagrow.

Over forty years have passed since the publication of this article by J.V. Mills, but no further comparative study has yet been published.

In 1973, at the 29th International Symposium of Oriental Studies, Paris, Michel Destombes announced the discovery of "A Chinese map drawn by Wang Pan (王泮) in the Ming Dynasty" in the Department of Oriental Manuscripts, B. Nationale. Scholars had long known of Wang Pan, who helped Matteo Ricci to compile and to print the first edition of the Chinese world map, but nobody had actually seen Wang Pan's map. For this reason, this essay, with a photocopy of Wang Pan's Map, generated great discussion among scholars as soon as it was published. Today, academic circles recognize that it is in fact a Chinese map drawn and supplemented by the Koreans on the basis of Wang Pan's *Yu di tu* (舆地图) in the early 17th century.

In 1974, an exhibition of Chinese and Japanese maps was held in the British Museum, London, which showed to the public 36 Chinese maps produced in different periods. Some remarkable maps among them were the Jesuits' Chinese maps and the *Qiankun wanguo quantu gujin renwu shiji* (乾坤万国全图 古今人物事迹) by Liang Zhou, "a comprehensive map (delineating) heaven and earth and the myriad countries and ancient and modern human affairs", from the Ming Dynasty, both from the private collection of Philip Robinson. Originally these maps were preserved in the library of the Jesuit College at Clermont, Paris. These maps purchased by Gerard Meerman, a Dutch collector, in 1764 when the college was disbanded. In 1824, they were bought from Meerman's heirs by Sir Thomas Phillipps, and then came into the possession of Philip Robinson. During this exhibition, Y. Jones, H. Nelson and H. Wallis jointly wrote an article describing the contents, date and features of these maps. Additionally, Nelson and P. Robinson wrote an essay outlining in detail the history of the maps and calling for a worldwide investigation of holdings of Chinese maps.

These two events in the early 70s' generated great interest in Chinese cartography. Cartographers and Sinologists in Europe, Japan and China produced articles introducing or proposing comparative studies of overseas Chinese maps. By means of this exchange between Eastern and Western scholars, a few previously unknown Chinese maps were located and some questions surrounding the sources of these maps were solved. Since the eighties, various exhibitions of Chinese maps have been held in Germany, France and Belgium.

Due to limited resources, Chinese scholars have had little knowledge about how many or which Chinese maps are preserved abroad. A small number of scholars from China have studied Chinese maps preserved in the United Kingdom, France, Spain and Japan. A few scholars in Japan, the United Kingdom, France, Germany and Italy have done similar work. Although most Western cartographers have strong interests in overseas Chinese maps and recognize the need to introduce them to the public in libraries and museums, they cannot engage in comprehensive surveys and research due to a lack of knowledge of the language, cartography and historical geography of China. Completed with the co-operation of scholars in China and overseas, the present work attempts to provide a relatively complete descriptive catalogue based on a survey of Chinese maps in European collections. Although it is primarily a work of a foundational character, it is hoped that it will serve as an aid to further research.

II. Survey and study of the author

II.1. Introduction

Since 1991, the author has visited libraries and museums in the Netherlands, Sweden, Denmark, Germany, Austria, France, Belgium, the United Kingdom, Italy and the Vatican and looked at over one thousand Chinese maps. Curators have generously permitted me to take photographs of their collections, allowing me to engage in comparative research. Here I would like to express my sincere appreciation to them. On the basis of the materials collected from overseas, I have compiled "A descriptive catalogue of pre-1900 Chinese maps seen in Europe", published in 1996, as the first volume of a proposed work entitled "Old Chinese maps overseas". I shall compile further volumes after future surveys and studies in other countries.

This is a documentary catalogue of pre-1900 Chinese maps overseas, with descriptions in both Chinese and English, arranging the maps chronologically within their category. In this book, the readers will find out information about old Chinese maps held in research libraries, and set them into a comparative framework to research.

Because modern techniques of surveying, mapping and printing spread throughout China from the end of the 19th century, and because the style of most Chinese maps tended to be consistent with European maps, there are quite a large number of printed maps available. However, during the same period, a few traditional woodcut maps were still being produced in private printing houses, although these were no longer the main form of Chinese cartography. As a result, maps produced after 1900 are not included in this catalogue.

Some of the Chinese maps were compiled by the Jesuits on the basis of the techniques of the European cartography and the information of the Chinese maps held in China. Because of their great value for research they are included in this catalogue, as well as the specific maps drawn by non-Chinese people, but in Chinese.

II.2. Determining the date of Chinese maps

For pre-1900 Chinese maps, what is most difficult to determine is the date of their production and the administrative situation of their depiction. Most maps in the Ming and Qing dynasties are without a mapping date or the name of the drawer or engraver. Many maps, including government maps, have also lost their title or have a postscript on a label stuck on the cover or verso of the map. Its date is, therefore, to be determined only by the contents of the map, and in this book I have used the following methods to do this:

1. *Administrative divisions*

First, a common and effective way to determine the mapping date is according to changes in the administrative units and divisions of China. I have laid out a table of changes in administrative unit: the names and administrative level of province, prefecture, sub-prefecture, county and military post. These are arranged chronologically as the basis for historical documents to be used for determining the date of the maps. One deficiency of this method is that, given the vast size of the teritory, it was very difficult for cartographers to gain the latest information on administrative units without official local chronicles. It was common that even when new units are designated in central areas, the old adminitrative situation is still depicted in the border areas on the same map. In other cases, old administrative divisions are depicted even when the form of the map, e.g. its style and colour, shows that it was produced in a later period. I am therefore only able to indicate an approximate mapping date based on the time of the adminitrative division depicted.

2. *Language*

Another means of determining a probable period of mapping is by refering to certain taboos of written language related to Chinese emperors. For example, if on a map all examples of the character *Ning* have been changed, lack a stroke or have been replaced by another character altogether, we can say that this map must have been drawn no earlier than 1821. This is because this is the first year of Emperor Daoguang (1821–1850) and the same character appears in his name, and its use elsewhere was considered taboo. The limitation of this method arrives from the different degree of rigour with which the taboo rule was applied in different periods. It was more rigorous in the later rather than in the earlier stages of the Ming Dynasty. In the Qing Dynasty it was most rigorously applied during the reigns of Yongzheng (1723–1735) and Qianlong (1736–1795), when most place names were changed in accordance with the taboos. After the reign of Xianfeng (1851–1861), the rule was gradually relaxed, and many maps made in the later Qing Dynasty often violated it. In one example, a taboo is both observed and violated on a map from a private press. This causes problems for accurate determination: e.g. *Qi sheng yan hai tu* (七省沿海图), "A Coastal map along the seven provinces".

3. *Historical geography*

If the above methods fail to determine the mapping date, we can turn to historical geography. Here I would like to introduce one example from my own research.

Several manuscripts showing the area of the lower Yellow River (黄河), Huai River (淮河) and the Grand Canal (大运河) are preserved separately in the British Library, the Royal Geographical Society, London, and National Geographical Society of Italy. They all have similar contents, are without titles and postscripts, the administrative units register no changes after 1800, and they have no character of taboo. It is therefore hard to determine the dates. However, by checking the lower course of the Yellow River from Qingjiangpu (清江浦) to the mouth, it is a apparent that one map shows it as a meander, two maps indicate cut-off places for the straighten line of the course and another two maps show two ox-bow lakes.

According to a historical document, a water conservancy project managed by Wu Jing (吴璥) was undertaken in 1803. Therefore, the map showing a meander must have been drawn before 1803, the two maps indicating the cut-off places can be dated to not long after the start of the project in 1803, and the two maps showing two ox-bow lakes are the most recent. This date can also be used for comparative dating of other maps.

4. *Library acquisition date*

Finally, library acquisition dates for Chinese maps are also of great value.

For example, a universal map of the Qianlong era was thought by Western scholars to have been drawn in 1743 because of a text on the map which states: "Qianshanzhai post established in the 8th year of Qianlong". This text is useful evidence, but this means that the map is no earlier than 1743 rather than that it was made at that time. So it would be a mistake to decide on an exact date for a map without a clear time engraved on it solely on this basis. Fortunately, a copy is preserved in the Bodleian Library, Oxford, with a register signed by Mr Frederick Pigon on August 1st, 1750. The register shows that this map was brought to England from Edo (Tokyo) in Japan. In view of shipping time, delays etc., I think that it may be more appropriate to determine the date of the map as between 1743 and 1749.

The register in the British Library, London is most notable in this regard, but not every library preserves such a strict and complete register, so determining the date of a map should be a comprehensive process which includes style, type and colour in addition to the above methods.

II.3. A survey of Chinese maps collection in Europe

The body of Chinese maps in European collections seen by the author up to the time of writting can be briefly summarized as follows.

Britain:

The British Library of London holds more than 200 maps, in both sheet and book form. These are divided between the Map Library, the Oriental and India Office Collections and the Department of Manuscripts. Most of these were donated by people visiting China from the 19th Century onward. More than 100 of these maps cover the period from the 1790s to the 1820s (i.e. from the reigns of the Jiaqing and Daoguang Emperors). These are mostly depictions of military posts and garrisons produced by the provincial governments of Jiangsu, Zhejiang, Fujian and Guangdong, and carry the seals of various levels of local government in both Manchu and Chinese.

These are official maps, unlikely to have circulated amongst ordinary civilians, so they are clearly booties taken away by British forces during the Opium Wars. One example is a map with 15 sheets of the counties and military posts of Ningbo Prefecture in Zhejiang province, presented by Hugh Gough, Commander of the British Expeditionary Forces in the First Opium War of 1840–1842. More than 50 military maps that were collected by John Robert Morrison, are in the Oriental and India Office Collections. There are also a number of rare items among the maps the library has purchased over the course of the last 50 years. These include a globe inscribed in Chinese that was made by the Jesuit missionaries Manuel Dias and Nicolo Longobard. Few examples of such globes can be found elsewhere in the world. The British Library has catalogued and arranged these maps extremely thoroughly, making it very convenient for visiting researchers to use them.

The Royal Geographical Society, Public Documents Office, and the Royal Asiatic Society of London also hold close to more than 100 Chinese maps between them, including maps of the south east coast of China and of Yunnan, Guizhou, Sichuan, Mongolia, Tibet and Xinjiang. Among these are maps of coast defences, waterways, administrative units and communication networks. A very large number of these were collected by the British missionary William Lockhart, who came to China in 1838 and worked as a doctor in Canton, Macao, Zhoushan and Shanghai. He gave these maps to the Royal Geographical Society on his return to Britain in 1858.

Over 10 Chinese maps are held in the libraries of Oxford and Cambridge universities, and a similar number are kept in the National Museum of Scotland in Edinburgh. Of these, the late 17th and early 18th century maps from the reign of the Kangxi emperor held at Oxford are among the rarest maps in the whole country. There are also some valuable maps held by private collectors, in particular maps from the late Ming that were dispersed after the dissolution of the Jesuit College at Clermont in Paris in the middle of the 18th century, which later fell into the hands of private collectors in Britain. These were put on public display in 1974. After sale by public auction in 1991, a number of these maps have now been taken to France and Sweden.

Overall, it can be said that Chinese maps held in Britain are not especially early ones, most dating from the middle of the 18th century (i.e. from the Qianlong and Jiaqing reigns onwards). This reflects the time period and historical context of British contacts with East Asia.

Germany:

Prior to German re-unification most Chinese maps in Germany were held separately in the Map and Oriental sections of the National Libraries of East and West Berlin. Among these are close to 100 maps of the counties in Shandong and Zhili (modern Hebei), dating from the second half of the 19th century to the beginning of the 20th century. These cover almost the entire area of the two provinces and include the prefectures, sub-prefectures and counties as well as the rivers and waterways of the region. Apart from this, there are also several maps of Heilongjiang, Jilin, Mukden, Fujian and Guangxi. These include specialised maps and maps of administrative areas, as well as long scrolls of the Great Wall, and of the noval garrisons and posts on the Yangtze River. After World War II, the books and maps of the German national library were divided between three places in East and West Germany. In the 1960s the West German collection was concentrated in the Staattsbibliothek Preussischer Kulturbestiz. Prior to the war, there was a handwritten catalogue of the maps of the German National Library, and this was held in the maps department of the Deutsche Staatsbibliothek of the former German Democratic Republic before 1992. However, following World War II some of the Chinese maps recorded in this catalogue were held in the same library while the remainder were housed in West Berlin, making it very inconvenient to examine them. The author was unable to see the entire collection during his visit. In addition to these holdings, the manuscript department of the Bayerische Staatsbibliothek in Munich houses long scroll maps of the Yellow River and of the naval defence posts along the Yangtze river in Anhui, as well as 8 maps of Licheng and Pingyao counties in Shanxi province. Most of these were made during the Xianfeng and Tongzhi reigns in the middle of the 19th century.

France:

The Maps section and Oriental Manuscripts section of the Bibliotheque Nationale in Paris holds over 40 hand-drawn and printed maps from the Ming and Qing dynasties. One of the most valuable of these is a large hand-coloured map on silk produced by Korean cartographers on the basis of the Yudi Tu ("A Map of Lands and Territories") annotated by Wang Pan, with additions and supplements. It probablly dates from between the 1600 and 1626 (i.e. between the Wanli and Tianqi reigns). Maps of this type are extremely rare. In addition, there are three large hand-painted silk wall maps of the Yellow River, Huai River and the Grand Canal on the border between Jiangsu and Shandong provinces dating from the 17th and 18th centuries.

Two large hanging scrolls of the Huguang region (Hubei and Hunan provinces) which date about the end of the 1790s (from the early Jiaqing reign). Both maps are seldom found in other collections.

The Bibliotheque Nationale has always put considerable effort into the acquisition of early printed Chinese maps such as the *Kun yu quan tu* ("A Complete Map of the World"), produced by the Jesuit Ferdinand Verbiest. The collection of different editions of the same map at this site is thus particularly complete. It has been reported that one or two Chinese maps made by the Jesuits were purchased by the Bibliotheque Nationale at an auction at Sotheby's in London in 1990.

A published catalogue produced by Morris Gulan in 1902 is in the Oriental Manuscript Department of the Bibliotheque Nationale. This incorporates the handwritten catalogue of the holdings of Chinese and Japanese maps done by Felix Pauly in 1892 in its section on geography. At present most of the Chinese maps are preserved in the library's maps section, where they are divided between special collections (Res.Ge) and the ordinary collection.

Vatican city:

Most of the maps of China produced by Jesuit missionaries before the middle of the Ming and the beginning of the Qing dynasties are now housed in the Vatican library, the Biblioteca Apostolica Vaticana. These include two copies of the 1602 edition of the *Kun yu wan guo quan tu* "Matteo Ricci's World Map" (a fragmentary copy of a Qing dynasty reprint is held in the Biblioteca Nazionale, Roma). The *Chi dao nan bei liang zong xing tu*, "Schall von Bell's Map of the Celestial Hemispheres of South and North on Equatorial co-ordinates", checked by Jacques Rho; the *Zhongguo zong yutu*, "A Map of China" by Michel Boym and the *Kunyu quantu*, "A World Map of Francesco Sambiasi". In the 1950s, M.J. Merjer researched a long coloured scroll of the Great Wall held in the collection of the Lateran Museum. That museum was closed in 1962, and later this scroll was moved to the Vatican Museum, and is now housed in the East Asian special collection of the Monumenti Musei e Gallerie Pontificie, Vatican City.

Italy:

An atlas of China made by Jesuit Michele Ruggieri based on the *Guangyu tu* but with place-names appended in Latin, which is held in the Archivio di Stato di Roma, has recently been republished. Two copies of the first copperplate edition as well as a subsequent edition with corrections and additions of the Kangxi's atlas on the basis of a survey carried out by Jesuit missionaries and Chinese officials are held in the Instituto Universitario Orientale in Napoli, and the Biblioteca Universitaria di Bologna and the Societa Geografica Italiana.

Of the over two hundred Ming and Qing manuscript and printed maps held in the Societa Geografica Italiana relatively rare items include a volume with sixteen hand-painted maps of the *Gan su quan zhen tu ce* (甘肃全镇图册), "An Atlas of the Complete Garrisons of Gansu in Ming dynasty", and an atlas with twenty-eight hand-painted maps with explanation of the garrison of Datong Defense Command, which is one of the nine frontiers along the Great Wall in Ming dynasty. A hand-drawn view of the Imperial Summer Residence in Chengde (Jehol) including the eight outlying temples; an atlas with seven hand-coloured maps of the *Jun xing liang yun shui lu cheng zhan li shu* (军行粮运水陆程站里数), "The Stations and Mileages of the Land and Water Transportation routes for the Army", from Guangxi province; a fifteen section Qing manuscript of the *Huang liang tai liu he quan tu shuo* (黄粮台六河全图说), "An Atlas with Explanation of the Six Rivers in the Prefecture of Yunnan", as well as long hand-coloured scrolls of the Yellow River, the Grand Canal, and the Hutuo River. There are also several dozen Qing official hand-coloured maps of the military posts, forts, garrisons along the coast, and over four hundred illustrations with explanatory texts of the non-Han areas of Guangdong, Guangxi, Guizhou and Yunnan, showing settlements, dress and customs etc. These maps were probably purchased by Giuseppe Ros. He worked as an Italian diplomat in China for more than thirty years in the first half of this century, and later donated them to the Societa Geografica Italiana.

In the absence of a new catalogue, the Chinese maps holdings of the Vatican library may be consulted using the typescript catalogue produced by Paul Pelliot in 1922. In the 1980s Professor Giuliano Bertuccioli of the Oriental Studies Department of Rome University compiled a typewritten catalogue of all Japanese and Chinese maps held by the Societa Geografica Italiana. The work of these scholars has been of great assistance to later researches.

Sweden:

A few Chinese maps are held in the Ethnografiska Museum in Stockholm. These were brought back from China by Sven Hedin and are mostly in the late Qing. There are also some holdings in the Kunglige Biblioteket, Stockholm and Universiteitsbiblioteket, Uppsala.

The author did not see the maps held by the private collector L.Bagrow, but his collections undoubtedly contain some valuable Chinese maps.

Denmark:

Chinese maps are concentrated in the Kongelige Bibliotek in Copenhagen. These are mostly printed maps of the whole country or of specific regions dating of the middle of 18th century onwards.

The Netherlands:

Apart from an early print of the *Guangyu tu* held in the Meermanno-Westreenianum Museum of The Hague, there is an anonymous hand-drawn map of the inner city of Peking in the same collection, and a copy of the *Kunyu quantu* by Fedinand Verbiest in the Maritiem Museum, Amsterdam. All Chinese maps are kept in the Sinologisch Instituut of Leiden University. The rarest maps in the Leiden collection are two folding maps painted on silk which depict the stations on the Grand Canal in the prefecture of Yangzhou in the mid 18th century (Qianlong Era). These are probably maps submitted to the throne by officials along the canal route to explain the resting palaces prior to the Qianlong emperor's fifth tour to Hangzhou.

Austria:

The map section of the Oesterreichische Nationalbibliothek of Wien holds more than ten Chinese maps. These include a copy of Ricci's World Map reprinted with alterations in the early Qing using the 1602 edition engraved by Li Zhizao. It was present to the Holy Roman Emperor Leopold I in 1672 by the Jesuit missionary Prospero Intorcetta who had been living in China. Although it is also an early Qing reprinting, it is somewhat different from the version hanging on the wall of the Royal Geographical Society in London.

Belgium:

Few Chinese maps are preserved in Belgium, and most of these are held in the Koninklijke Bibliotheek van Belgie, Brussels. Most of them date from the mid and late Qing dynasty.

At the Katholieke Universiteit Leuven there is a Ferdinand Verbiest Project by Prof. Dr. U. Libbrecht which researchs Verbiest and other Jesuits, but most of the maps collection there consists of reproductions.

III. Academic value of chinese maps found overseas

III.1. How to comprehend hand-drawn and printed maps

The historical circumstances of the acquisition of Chinese maps by European are interestingly connected to the relationship between printed and hand-drawn maps. While overseas the author was often asked if manuscript maps are the most valuable. Overseas libraries and museums all seem to consider hand-drawn maps as the most valuable, and appear to believe that because a printed map can be reproduced many times, it circulates more widely. On the other hand hand-drawn maps are only drawn once, so they are thought to be unique. In fact the situation is somewhat more complex. The author saw many copies of long coloured scrolls depicting the Chinese coastline or the Yellow River. These scrolls are often painted on silk or paper, with fine silken backing, and are very exquisite. They seem to be items offered to the emperor as a tribute. However, if closely examined, many are the work of artists from the later period. In this regard the maps of the coastline held in London and Rome are worth analyzing.

These maps represent pictorially the Chinese coast of the seven provinces from Korea to Annam along the length of the scroll. At the beginning of each scroll, there is usually a map of the Eastern Hemisphere. Following the map of the coast, there are maps of Hainan Island, the Pescadores and Taiwan Island. The copies of these maps held overseas are all basically the same, only differing in the amount of notes. The ancestor of these maps is probably the fold map edited by Chen Lunjiong in 1730. There are no written notes on Chen's maps, however from the 1790s onwards non-official copies appeared continuously. They generally all have written notes, describing water conditions, sand bars, marine defences, and wind directions on the shipping routes. These non-official maps have a common feature:

One, they depict frequently the administrative divisions existing in 1730 and do not give later changes. The sub-prefecture (Ting) established in the Qianlong and Jiaqing reigns is not mentioned, so the maps do not depict the situation of the times of their production and easily give the mistaken impression of having been painted earlier than they actually were.

Secondly, imperial name taboo rules are not strictly observed and mistaken or missing characters are common. On the same map taboo characters may sometimes be changed and sometimes be left as they are. Also, missing characters make some sentences unintelligible. For example in one of the notes on Lianzhou (廉州) in Guangdong province, it says "if sailing from Hai'an to Lianzhou, the south wind is favourable to ships while the north wind is not favourable". The copies of London and Rome all say "the south wind is favourable the north wind is favorable". The absence of a "not" makes the sentence difficult to be comprehended. One suspects that this is a mistake of a later copyist.

Thirdly, on the some of the maps the treaty ports established after 1861 are added and the mouth of the Yellow River is always placed in northern Jiangsu.

The appearance of these non-official maps and their spreading overseas, on the one hand reflects how the maritime border of China was coming under threat in the mid-19th century, creating a need for maps in order to strengthen coastal defence. These maps could also reflect Westerners increasing need for Chinese maps, stimulating the growth of an non-official map drawing enterprise to profit from this demand. If this is so, these hand drawn maps are not necessarily more valuable than printed maps. The most important issue is to evaluate the time of the production of the map and its origin.

III.2. Academic value of chinese maps found overseas

What is the academic value of these old overseas Chinese maps?

1. Overseas Chinese maps provide a wealth of real data for the study of the history of Chinese cartography.

Books on the history of Chinese cartography are unable to describe completely the maps produced from the Yuan Dynasty to the first half of the Ming Dynasty because most of the maps recorded in historical sources have been lost. Overseas Chinese maps offer valuable material to help to overcome this gap, although they are mainly from the late Ming and after.

These maps reveal a style of map from the Ming period different from the cartographic system of the *Guang yu tu* ("An Atlas with text of China"). From the relationships of origin between these maps, we can trace the evolution of the major elements and modes of presentation of traditional Chinese map-making, in particular for the general maps of China. Few copies of these maps are preserved in China, so research done on them has filled some of the gaps in our knowledge of the development of Chinese cartography.

At this point it is important to mention the role of Korea, a close neighbour of China. Old Korean maps and Chinese maps taken there during the Yuan and Ming Dynasties greatly affected the development of cartography in the Ming and Qing Dynasties. Examples include the copies of Wang Pan's map and a general map of the Ming Empire *Hun yi jiang li li dai guo du zhi tu* (混一疆里历代国都之图), "Comprehensive Map of the World and the Nation's Successive Capitals" made by Yi Hoe (李荟) and Kwon Kun (权近) in 1402, on the basis of Qing Jun's map *Hun Yi Jiang Li Tu* (清浚：混一疆里图) and Li Zemin's map *Sheng Jiao Guang Bei Tu* (李泽民：声教广被图) that were brought to Korea by Envoy Kim Sa-hyong (金士衡). No copies of the general maps in the Yuan Dynasty were kept in China, and only one copy of the general maps, a map *Yu Di Tu* with a postscript by Yang Ziqi (杨子器题跋: 舆地图) produced before 1526 is preserved in Dalian (大连). Therefore, understanding of the continuity and change in the shape, mapping rules and symbolic styles of the general maps in China for more than two centuries, from the second half of the thirteenth century to the beginning of the sixteenth century depends on maps drawn by Korean cartographers.

2. Overseas Chinese maps reflect scientific and cultural exchanges between China and foreign countries extending over a period of several centuries.

From their inception in the 15th century, depictions of the geography and landform of East Asia and China on World maps by European cartographers show a gradual increase in detail. Correct understanding of the geography of Asia by the West resulted partly from observations and surveys by Western visitors, but most information was drawn from traditional Chinese maps brought to the West, especially before the middle of the 19th century.

For example, the Chinese coast on the maps of Asia printed by Europeans in the 17th century, such as Gerard Mercator in Duisburg (1569), Ludovico Georgio in Antwerp (1584), Jodocus Hondius in Amsterdam (1606) and Willem Blaeu in Amsterdam (1618), was drawn almost as a straight line. All the rivers in China were depicted as linked up to each other. These earlier maps clearly lack exactitude and were drawn neither on the basis of Western surveys nor on Chinese maps. Furthermore, even though the Dutch voyaged along the western coast of Taiwan Island and surveyed the area, they still drew Taiwan Island as three small islands on their maps because the Dutch regarded the three estuaries as channels.

Until the Jesuits Matteo Ricci (利玛窦 1552～1610), Michele Ruggieri (罗明坚 1543～1607), Alvarez de Samedo (曾德昭 1585～1658), Michael Boym (卜弥格 1612～1659), Martino Martini (卫匡国 1614～1661) and

Philippus Couplet (柏应理 1622～1693) sent their maps, based on Chinese maps to Europe, the “Map of China” drawn by European cartographers, never conformed to geographic reality. Martino Martini's “Novus Atlas Sinensis” (中国新图志), published by Joan Blaeu in Amsterdam in 1655, actually became a model for the Map of China in Western publishing houses and caused Father Martino Martini to be called “the Father of Chinese Geography” in Europe. In fact, Father Martino Martini's “Novus Atlas Sinensis” drew heavily on Luo Hongxian's *Guang yu tu* and other Chinese maps, something Martini acknowledged in the Preface of his atlas. As for the maps of Alvarez de Samedo and Michael Boym, they represent a reproduction of the Universal Map of the Ming Empire in 1605, (*bei zhi huang ming yi tong xing shi fen yie ren wu chu chu quan lan* 备志皇明一统形势 分野人物出处全览) as preserved in the Biblioteka Czartoryskich in Cracov, Poland. This article does not intend to deny the contribution of the Jesuits in China to cultural exchange between China and Europe, but it is clear that the information gained from traditional Chinese maps favoured the Jesuits' achievements in cartography. Ricci's world map helped make the East meet the West, and overseas Chinese maps have helped make the West understand the East.

The next change to the face of the “Map of China” in Europe happened in the middle of the 18th Century when sketches of the general map of Emperor Kangxi (*Kang xi huang yu quan lan tu* 康熙皇舆全览图) were carried to Paris for reproduction on copper plates. Jean Baptiste d'Anville (唐维尔) took advantage of the occasion to compile his “Nouvelle Atlas de la Chine, de la Tartars Chinosis et du Thibet” (中国、蒙古与西藏新地图集) which was published in Paris in 1737. D'Anville's atlas gave greater and more accurate detail and became a new model for the “Map of China” in Europe.

Thus the exchange of maps between China and foreign countries played a very important role in enriching human knowledge of the Earth, Geography and Cartography.

3. The specialized maps drawn or engraved at different historical periods, such as those of China's border administration, coastline, the Great Wall, urban plans and military posts offer a relatively accurate picture of a specific time and place. This can be used to research the historical, political or military geography of particular regions. Since the majority of these specialized maps were produced by those directly involved with the region in question, they constitute a valuable primary historical source.

Different maps use different means to represent geography. The direction of perspective and the amount of detail reflect both the scientific level and the thought and culture of the period. By analyzing these specialized maps we can see how Chinese people at that time understood a specific piece of geographic space, and how they used the medium of maps to express their objectives, that is not only to understand the space themselves but also to make it understandable to others.

For example, coastal maps of China are often drawn with visual direction, that is different from the standardized orientation used in modern cartography. One type is represented by the "Map of the Coastline in the 7 Provinces", which follows the coastline always oriented from the sea towards the land. Another type is the map of maritime defence posts which illustrates the defensive situation on a section of coastline. These generally adopt on orientation from the land towards the sea. The former appears to be for use by ships sailing along the coast, while the later was used by army forces stationed on land. As a result, the two required a different viewpoint causing the orientation of the maps to differ. However, they all use the same pictorial illustration reflecting how traditional Chinese maps were born out of an original tradition of "Maps of Perceived Terrain", closely connected to landscape painting.

In the Qing dynasty, there was a great increase in the number of maps depicting the entire course of the Grand Canal and the Yellow River, reflecting the country concern with controlling these waterways to ensure the transportation of grain. These maps show the water control works undertaken at various periods to control and safeguard both the Yellow River and the Grand Canal. They also depict the stages in the evolution of the course of the Yellow River. Following the diversion of the Yellow River into the Huai Basin, there was a progression from a natural meander to the cutting off of the meander and finally to the emergence of a new course and an ox-bow lake. This helps us to understand the hydrological history of the Yellow River more clearly.

The legends on the map of the river defences on the Yangtze note the distances between the banks, the positions of sand-bars and their lengths, while the maps of coastline show the shape of the river mouth, the positions of sandbanks, and their breadth and depth. These provide comparative statistics with which to research the rate and tendency of erosion and sedimentation on both banks of the Yangtze River over the last 100 years or so.

Maps of the borders and of the roads from different periods reflect the process of China's expansion in border regions, as well as the position and direction of the national boundaries, together with the changes in routes of communication and their distribution. These are valuable materials for academic research and also have practical value.

Urban maps reveal the layout of streets and buildings in Chinese cities in different periods. The spatial and temporal information that these maps give about the evolution of cities is an indispensable primary source. They can be used for research on urban historical geography and can also aid modern urban management in the utilization and preservation of the traditional urban cultural landscape.

Of course, the above-mentioned Chinese maps held overseas are not the only ones of academic value. Maps held within China also have great practical and scholar importance. However, they have not been generally recognized by scholars and the public until now.

Overall, the surveying and researching of Chinese maps in overseas collections is a task of great significance, the importance of which has only recently been recognized. The success of this work depends on international co-operation. The author expects that this book helps making Chinese maps overseas accessible and contributes to bring together academic efforts on this subject.

REFERENCES

Baddeley, J.F., Another Jesuit World Map made in China, *The Geographical Journal* 53.1, 1919.

Baddeley, J.F., Father Matteo Ricci's Chinese World Maps, *The Geographical Journal* 50.4, 1917.

Bertuccioli, G. (白佐良), Per una Storia della Sinologia Italiana: Prime Note su Alcuni Sinologhi e Interpreti di Cinese. 74, *Mondo Cinese*, XIX.2, 1991, Milano.

Bjorck & Borjesson, Catalogue 522, *Bibliotheca Asiatica*, Part 1. Stockholm, 1990.

D'Elia, P.M., Recent discoveries and new studies (1938-1960) on the world map in Chinese of Father Matteo Ricci, *Monumenta Serica* 20, 1961.

Destombes, M., Une carte chinoise du XVI siècle découverte à la Bibliothèque Nationale, *Journal Asiatique* CCLXII, 1974, 193-212.

Fuchs, W., *Der Jesuiten-Atlas der K'anghsi-Zeit: China und die Aussenlander*. Peiping, 1941.

Fuchs, W., *Der Jesuiten-Atlas der Khang-Hsi Zeit, seine Entstehungsgeschichte nebst Namenindices Fur die Karten der Mandjurei, Mongolei, Ostturkestan und Tibet, mit Wiedergabe der Jesuiten-Karten in Originalgrosse*. Fu-Jen Univ. Press, Peiping, 1943.

Fuchs, W., *Chinesische Handschriften und Mandjurische Handschriften und seltene Drucke: nebst einer Standortliste der sonstigen Madjurica*. Wiesbaden, 1966.

Giles, L., Translations from the Chinese World Map of Father Ricci, *Geographical Journal* 52.6, 1918; 53.1, 1919.

Harley, J.B., and Woodward, D.A., *The history of cartography* 2, 1994.

Heawood, H., The Relationships of the Ricci Maps, *The Geographical Journal* 50, 1917.

Hirotada Kawamura, The Austrian National Library's Matteo Ricci World-map, *Human Geography* 40.5, 1988.

Hodson, Y., Map of the Orient at the British Library, *The Map Collector* 28, 1984.

Istituto Poligrafico e Zecca dello Stato, *Atlante della Cina*. Roma, 1993.

Mills, J.V., Chinese Coastal Maps, *Imago Mundi* 11, 1954, pp.151-168.

Mills, J.V., Three Chinese Maps, *British Museum Quarterly* XVIII.3, 1953, pp.65-67.

Meijer, M.J. A Map of the Great Wall of China, *Imago Mundi* 13, 1955, pp.110-115.

Needham, J., *Science & Civilisation in China*. Vol. III, The Sciences of the Earth, Cambrige Univ. Press, 1959.

Nelson, H., Maps from old Cathay, *Geographical Magazine* 47.11, 1975, pp.702-711.

Jones, Y., Nelson, H., and Wallis, H., *Chinese and Japanese Maps: an exhibition organised by the British Library at the British Museum February 1 – December 31, 1974*. London, 1974.

Wallis, H.M., and Grinstead, E.D., A Chinese terrestrial globe, A.D. 1623, *British Museum Quartely* 25, 1962, 83-91.

Wallis, H.M., The influence of Father Ricci on Far Eastern cartography, *Imago Mundi* 19, 1965, 38-45.

Wallis, H.M., Missionary cartographers to China, *Geographical Magazine* 47.12, 1975.

Wallis, H.M., Chinese Maps and Globes in the British Library and the Phillips Collection, *Chinese Studies*: papers presented at a colloquium at the SOAS, University of London, August, 1987.

Walravens, Hartmut, Father Verbiest's Chinese World Map (1674), *Imago Mundi* 43, 1991, 31-47.

Zogner, L., *China cartografica: Chinesische Kartenschätze und Europäische Forschungsdokumente*, Berlin, 1983.

Zurcher, E., Standaert, N.S.J., and Dudink, A., (ed.) *Bibliography of the Jesuit Mission in China ca. 1580 – ca. 1680*. Leiden, 1991.

WESTERN KNOWLEDGE OF GEOGRAPHY REFLECTED IN JUAN COBO'S *SHILU* 实录 (1593)*

LIU Dun
Institute for the History of Natural Science, Chinese Academy of Sciences, Beijing

1. Fray Juan Cobo and his writings

Little is known about Juan Cobo's early life except for the uncertain year of his birth, 1546, and the name of his hometown, Consuegra (now in Toledo Province, Spain). Juan Cobo (later Chinese name 高母羡) was sent to a school run by the Dominican Order in his early teens, after which he became a Friar at the monastery of Santo Tomás in Avila, where he stayed until he decided to leave Spain as a missionary to the Far East.

Like most Spanish Friars who went to the Orient in the sixteenth-century, Juan Cobo first went to Mexico where he spent several months before travelling on to the Philippines. When he arrived there in May of 1588, he settled in Manila and soon began his missionary work. But his primary objective was to enter China, the mysterious and vast continent to the north. Therefore, Juan Cobo immediately began to study Chinese with local Chinese immigrants in Manila. He studied dilligently, learned quickly, and not long thereafter could speak to local people, and could read and write about 3,000 Chinese characters. He was the first missionary in the Philippines who preached to local believers in Chinese, although he did so with some kind of Southern Fujian dialect (Hokkien).

His success was noted by Santiago Vera, the Governor of Manila, who attended one of Juan Cobo's sermons and was both very pleased and astonished to see that the Chinese congregation was so attracted by this Spanish Fray's preaching.

---

* The author is grateful to Dr. José Antonio Cervera for providing him with a copy of Fidel Villarroel's edition of Juan Cobo's *Shilu* [1], the main source upon which this paper is based. He is also indebted to Dr. Catherine Jami and Prof. Joseph Dauben for their help in correcting the English and for their useful comments as well.

In fact, Juan Cobo himself was greatly impressed by the Chinese people and their culture. In a letter to a friend he wrote:

> *"There is something that should be emphasized. The people we know and who come here are the lowest — sailors, fishermen, and handicraftsmen. They come for finding food, although they do not like doing this. It is amazing that they are so witty and so clever that we could not find even one who could not be spoken to, not about fishing but about letters, celestial movements, morals, courtesy, and justice. They know lots of moral philosophy, but without science."* [1]

Although we do not know with whom Juan Cobo studied Chinese literature, from the elegant sentences and various especially pertinent phrases from ancient Chinese classics which he quoted in his books written in Chinese, it may be assumed that there was at least one Chinese scholar working with him.

In 1592, Toyotomi Hideyoshi 丰臣秀吉 (1537-1598), the feudal lord and shogun who had just completed the unification of Japan and was launching the first war to invade Korea, sent an emissary to the Philippines. In return, the Spanish Governor, Pedro Gomez Dasmarinas, appointed Juan Cobo as his ambassador to Japan. Cobo went, and although he apparently made a success of his mission, he never returned to the Philippines. Unfortunately, his boat sank in a terrible storm on his way back to Manila in November, 1592.

With this brief sketch of Juan Cobo's life and career as a missionary in mind, it is now possible to consider his writings in more detail. In addition to some documents that may be attributed to him, it is known with certainty that he wrote or translated at least the following four books:

(1) ***Carta de la China*** (*Letter about China*)

This is a book of in the form of a letter dealing with daily life and customs of the Chinese living in Manila.

(2) ***Doctrina Cristiana en Letra y Lengua China***
(*Christian Doctrine in the Chinese Language*)

This is one of the first three books printed in the Philippines, in which Cobo introduces the basic doctrines of Christianity in the Chinese language[2].

---

[1] Cobo, *Carta a los religiosos*, see [1], p.60.

[2] The other two are the *Shilu*, also by Juan Cobo, as discussed in greater detail below; and the *Doctrina Christiana en Lengua Espanǒla y Tagala*, written by Dominicans who have also settled in the Philippines.

There is a copy in the Library of Congress, Washington D.C.; another copy is preserved in the Vatican Library.

(3) ***Espejo Rico del Claro Corazón*** (*Precious Mirror of the Clear Heart*), or, in pin yin, *Mingxin Baojian* 明心宝鉴 [3].

This is a handwritten, bilingual work: Chinese text in the left, while Spanish translation by Cobo in the right. This book contains many moral aphorisms and sentences selected from both Confucian and Taoist works, and Cobo himself mentions the original author as Lipo-Pun Huan. According to Paul Pelliot (1878-1945), the author of the *Mingxin Baojian* is Fan Liben 范立本, and this is the earliest work that translated ancient Chinese philosophy into a Western language[4].

(4) ***Shilu***

In what follows, Juan Cobo's *Shilu* will be the focus of more detailed analysis.

Fig. 1 – Portrait of Juan Cobo.

Fig. 2 – The cover page of the *Shilu*.

[3] I am grateful to Dr. Adrian Dudink for providing me several references about the *Mingxin Baojian*; among them, Wang Chongmin attributed it to a Taoist work compiled in the Song-Yuan periods. See [3].

[4] Pelliot did not give further information about the author except his styled name, Congdao 从道, and his hometown, Hangzhou 杭州. Apparently Lip-Pun Huan is the name of Fan Liben pronounced in Southern Fujian dialect. See [4]. I am indebted to Dr. Adrian Dudink for his help in confirming this note.

2. Juan Cobo's *Shilu* 实录

The complete title of the *Shilu* is *Bian Zhengjiao Zhenchuan Shilu* 辨正教真传实录, or in Spanish *Apologia de la Verdadera Religión en Letra y Lengua China* (Apology of the True Religion in Chinese Writing and Language).

As a matter of a fact, the *Shilu* is really a catechism, a manual of religion arranged as usual in the form of questions and answers meant to instruct believers, to win converts, and to teach the truth of Christianity. Somewhat earlier, in 1584, Michele Ruggieri 罗明坚 (1543-1607), an Italian Jesuit who was the first Western missionary to arrive in China in the sixteenth century, published his *Tianzhu Shilu* 天主实录 (Veritable Record of the Lord of Heaven). It is considered to be the first book about *xixue* 西学 (Western learning) written in Chinese and printed in China. Somewhat later, in 1604, the well-known Italian Jesuit Matteo Ricci 利玛窦 (1552-1610) published his most famous work, Tianzhu Shiyi 天主实义 (The True Meaning of the Lord of Heaven), a book that was quite popular among early seventeenth-century learned Chinese[5]. Thus Juan Cobo's *Shilu*, when it was published in Manila in the year 1593, appeared nine years later than Ruggieri's book but eleven years before Ricci's *Tianzhu Shiyi* was published in China.

It seems, in fact, that Cobo knew of Ruggieri's *Shilu*. On the other hand, it is not so clear whether Ricci had read Cobo's *Shilu* or not. In any case, the contents of Ruggieri's book are purely religious, whereas both Cobo's *Shilu* and Ricci's *Shiyi* combine, to some extent, religion with philosophy and science. Therefore Cobo's was the first book to appear in China, in Chinese, with a certain extent of scientific content, and consequently it is worth studying further the possible influence of Cobo's work on Matteo Ricci's proselytization methods in China.

All three books are also constructed in the form of a catechism, a device initiated by the Church Fathers in medieval times, and which became much more important after the second invention of printing in fifteenth-century Europe. In fact, the content of Juan Cobo's *Shilu* is a conversation between a Western missionary and a Chinese scholar. The former is referred to as *seng* 僧 or *sengshi* 僧师 (monk or master monk), whereas the latter is referred to as *xuezhe* 学者 (scholar). Their dialogues are full of wisdom, courtesy and mutual respect. There is no reason to doubt that the "monk" is Juan Cobo himself.

---

[5] The original title was *Tianxue Shiyi* 天学实义, which Matteo Ricci himself changed in later editions.

Except for a few lines of Spanish at the very beginning of the *Shilu*, the remainder of the text is in Chinese.

It is interesting to compare the *Shilu* with another book by Cobo, *Doctrina Cristiana en Letra y Lengua China.* The latter was written for ordinary Chinese Christians by simply presenting the basic ideas of Christianity. On the other hand, Cobo's *Shilu* was intended for those Chinese who were well-educated in traditional literature but wanted to know more about the origin and principles of the world, or as Cobo believed, the truth behind Christianity. Therefore, he used scientific knowledge and logical reasoning to make this truth as apparent a possible. In all, the *Shilu* consists of the following nine chapters:

1. Discussion of proofs of the true religion;
2. On the existence of an infinite being, the principle of all things;
3. Talking about infinite things;
4. On matters of geography;
5. About the reality of earthly things;
6. On plants of the earth and other vegetables;
7. On things of the animal kingdom;
8. On how animals know what they should eat and drink;
9. On how animals of the world know the medicines they must take.

In general, the first three chapters are more theological than philosophical, whereas the other six are philosophical and even scientific. It is likely that Juan Cobo never actually finished this book, due to his unexpected death in 1592.

Unlike Ricci's *Shiyi*, Cobo's *Shilu* was not widely circulated among learned Chinese, and there is no evidence that this book ever reached the mainland or was circulated in China. In fact, the only extant copy is kept in the National Library of Madrid, and only came to wider attention in 1986, when a Spanish Dominican, Fidel Villarroel, in collaboration with other two Dominicans working at the University of Santo Tomás in Manila, published a facsimile reproduction of the copy in Madrid, along with both Spanish and English introductions and translations.

3. Western knowledge of geography reflected in the *Shilu*

Although the ultimate goal of the *Shilu* was to propagate the Christian faith, Cobo placed particular emphasis on geography at the end of Chapter Four:

> *"The earth exists and fields exist; because there are fields, there are men; because there are men, there is wealth; because there is wealth, it has been used. Therefore it can be seen that geography was created by the Lord of Heaven. So the Lord of Heaven is the master of geography, and geography is the master of men, because men have to rely on geography."* [6]

Clearly there was important information about European ideas concerning geography in Cobo's *Shilu*, all of which was new to sixteenth and seventeenth-century learned Chinese.

3.1. Ptolemaic cosmology

Ptolemaic cosmology basically involves a geocentric system which was regarded by the medieval Church as a literally true description of the universe. Although the *Shilu* was written nearly half a century after Nicholas Copernicus (1473-1543) published his *De Revolutionibus* (1543), Juan Cobo did not mention the new Copernican picture of the universe.[7]

As for the traditional Aristotelian-Ptolemaic picture of the universe, although the number of spheres sometimes varied, the basic framework was always the same, with the central earth as the point from which all motions were defined according to Aristotelian principles.[8] In an illustration given

---

[6] 有地此有土，有土此有人，有人此有财，有财此有用也。是可见地理造作于天主，故天主者地理之主宰；世人运用于地理，故地理者世人之主宰也。

[7] Actually, we do not know whether Juan Cobo even knew about this new cosmology or not, although some authors argue that Cobo had studied astronomy. For example, Villarroel refers that Cobo might use Alexandro Piccolomini's *Della Sfera del Mondo* (1540) , see [1], p. 87; [2], p. 197.

[8] According to Aristotle, earth and water are heavy elements, and they can be at rest only when they are at the center of the cosmos, while air and fire have a tendency to rise, so their proper spheres are above the earth. Moreover, the heavenly bodies which were presumably made of a fifth element called "ether" were taken to move around a centrally-positioned earth. As Steven Shapin has pointed out in his *The Scientific Revolution* (Chicago, 1996), "the cosmos thus spun about the earth, the place where human beings lived, and in just that sense pre-Copernican cosmology was literally *anthropocentric*", pp. 22-24.

in the *Shilu*, the spheres around the earth are depicted in the following order: air, fire, the Moon, Mercury, Venus, the Sun, Mars, Jupiter, Saturn, fixed stars, and finally, the Lord of Heaven. It is this last region, beyond the fixed stars, where all the angels and just men see God.

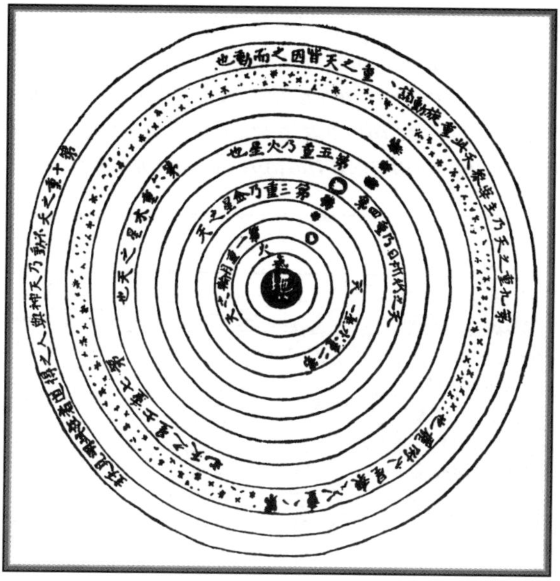

Fig. 3 – Ten Heavens.

3.2. Spherical Earth

The core of Chapter Four of the *Shilu* is devoted to arguments meant to establish that the earth is indeed a sphere. Although some thinkers had proposed similar ideas in ancient China, Juan Cobo was the first to present a systematic argument in Chinese on the spherical shape of the earth. For example, he offers the following arguments to support this theory.

> *"(1) Suppose two ships are on the sea at a distance of somewhat more than forty* li 里*; we can only see the mast of the other ship, while its body is not seen by our eyes. It is not that the ship is small or big, but what happens on land also happens at sea. If the land is round, the water is round also. And because the water is round,*

therefore a ship looms larger as it approaches floating on the water, and a man looking at her sees her growing bigger." [9]

"(2) If the form of the earth were square, it would naturally have four angles (like a vertical cliff). Hence (in one direction) a man could see a fire on a mountain from even one hundred li *away. Nevertheless, (in another direction) a man could not see a brilliant fire on the top of Tai mountain (a very high mountain), even if he were less than five* li *away from the fire. This is not due to the fact that the fire is big or small, but because... the fire on the mountain was hidden from the man even though he was near it.*" [10]

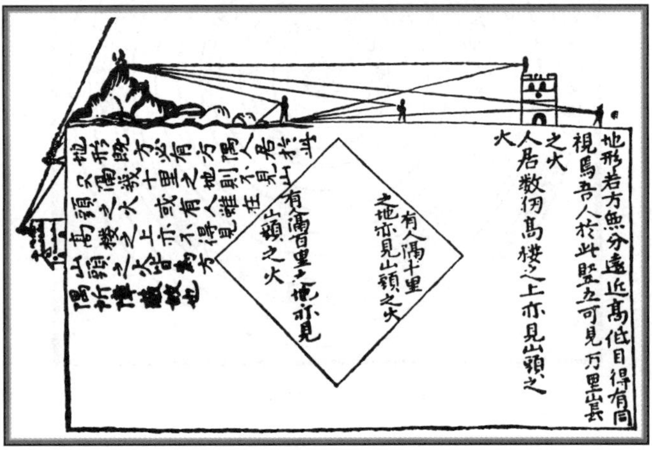

Fig 4 – If the form of earth were square.

"(3) The dim glow of the moon borrows its light from the sun, and is interposed by the earth... looking up the moon, the shadow of the earth can be seen as a segment of a circle. ... At any given time the eclipse of the moon is round but not square." [11]

---

[9] 于洋中两舟相遇，离有四十里之远，只见舡桅而已，舡身渺小不见；及至五里间，舡身则在吾目中矣。非舡之能小能大也，盖以水由地中行，地既圆活，水也圆活，水圆活，故舡之行由远及近，人之见舡者由小而能大也。

[10] 使地形而果方也，截然四正，头角分明，则明见百里之外无遗光矣，况山火何难见耶？然又有一人步不离于五里，乃泰山之颠爝火之光，反不能目见此火。非火之能小能大也，盖…火之在山者，近有所蔽；人之见火者，近有所遗也。

[11] 月色微光，借日之光以为光，且为地影冲障其中。…仰观月色分明，中有地之影，如弓之圆者在焉。…每时月蚀每圆而不方。

3.3. Circumference of the Earth

The circumference of the earth as computed by Cobo is 6300 *pu* 舖, where one *pu* equals ten *li* . He explains as follows:

> *"To know accurately the size of the earth, one must consider what the ancient scholars had used as tools. There are a total of 360 degrees around Heaven. The whole form of Heaven is divided into four quarters, and each quarter is equal to 90 degrees. ... Using a string (attached to a protractor) to measure the North pole at a place on ground, (then go southward) 175* li *from the place and measure the North pole to get 1 degree (as the difference). How can this be proved? Suppose a man stood in the North so that the North pole was just above his head; once he walked 175* li *and then measured the North pole by a string (attached to a protractor), the difference would be 1 degree. In the same way, the degrees would increase or decrease as he walked northward or southward. Therefore the size of the earth can be known. A quarter of (the circumference) of the earth equals to 1575* pu, *and the whole circumference is 6300* pu.*"* [12]

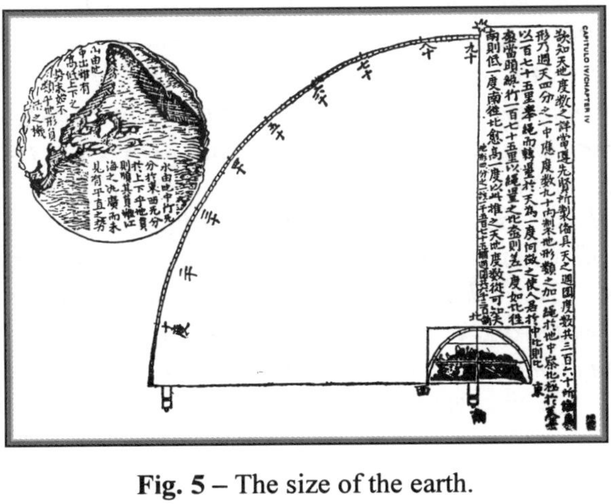

Fig. 5 – The size of the earth.

[12] 欲知天地度数之详，当尊先贤所制备具。天之圆周度数共三百六十，所备具□形乃周天四分之一中，应度数九十内，制地形类之。加一绳于地中，察北极于天上，地以一百五十七里举绳转量于天为一度。何以徵之？使人居于中北，则北极当头，才行一百五十七里，以绳量之北极则差一度。如北往南则低一度，以此推之，天地度数从可知矣。地形四分之一该一千五百七十五铺，周围共六千三百铺。

That is

$$176 \times 90 = 1575\ (pu)$$
$$1575 \times 4 = 6300\ (pu)$$
$$6300 \times 10 = 63000\ (li)$$

Unfortunately, we do not know where this data came from, nor what scale Juan Cobo used, so it is difficult to estimate the accuracy of his conclusions. However, this is the first text written in Chinese which directly states the size of the earth based on its roundness and the calculations which follow from this assumption.

3.4. Different climate zones

The text says:

"Under the universe there are six different divisions, and each one is subdivided into three parts. Three parts are near the North Pole and the other three parts are near the South Pole. But in the lands near the North Pole and South Pole, the four seasons are not defined. Among (the six divisions), there are two in which the climate is very cold. Even though one puts on thick furs, one could not defend oneself. Men cannot inhabit these cold lands. There are two in which the climate is very hot, but who can (survive) without heat? ... Therefore those who can bear heat can inhabit these lands. As for the other two divisions of the climate, the heat and the cold may be harmonized. To drink hot soup and to wear warm fur in winter, to drink cold water and to wear hemp clothes in summer; when spring and autumn come, the weather is pleasant and suitable." [13]

[13] 六合之下，别为六区；析而论之，六区之间，各分三分。对北极者三分，对南极者亦三分也。然近南北之地，四时无定。其中二分之气甚冷，虽重裘救寒不能止其寒矣，世人不能居斯冷地也。二分之气虽热，谁能去热？…故可以制其热者也可以居其地。至于二分之气冷热相和，冬时冷则饮汤矣，冬裘矣，夏时热则饮水矣，夏葛矣。及时之春时之秋，暘和宣畅，温凉得宜。

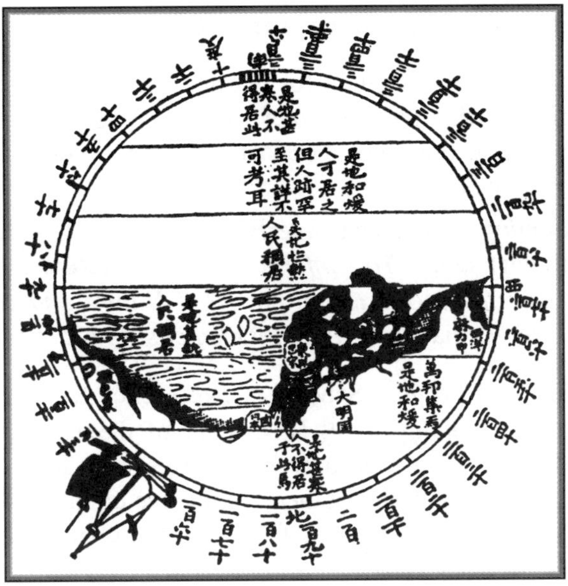

Fig. 6 – Climate zones and the world map.

3.5. World Map

In addition to Juan Cobo's description of differences between climate zones, the *Shilu* also supplies a map of the world, although it is only a very rough one, and only covers the Northern Hemisphere. Among places marked in particular on this map: *Daming guo* 大明国 (China), Weiseguo 微色果 (Mexico), *Riben guo* 日本国 (Japan), Lü song 吕宋 (Luzon), Malijia 麻力甲 (Malacca). All these places were clearly of special importance for the sixteenth-century Spanish who came to the Far East. Meanwhile, as regards latitude, the locations of all these places as given on the map in the *Shilu* are basically correct. For example, Mexico, Japan, Luzon, and Malacca are placed, respectively, at about 25° – 30°, 15° – 30°, 20°, and 5° north latitude. On the other hand, longitudes were not included, but despite this shortcoming, Juan Cobo's map is an accurate reflection of the level of mathematical and geographical knowledge in his day.

As for the Southern Hemisphere, Juan Cobo only says:

"There are very few vestiges of human beings (there)*, so it is not possible to examine the details."* [14]

---

[14] 人迹罕至，其详不可考耳

4. The character and significance of the *Shilu*

Because the *Shilu* was written for learned Chinese, there are numerous phrases and idioms which were either directly quoted or slightly altered from Chinese classical literature. For instance, at the very beginning of the book, Juan Cobo writes:

> *"The Guangming Xiangsheng Xuezhe* 光明先圣学者 (The Brilliant Prophet Scholar) *said: "To conform with one's nature is called* dao 道, *to cultivate the* dao *is called* jiao 教." *Nature and* dao *are the same, how can the* jiao *have different methods (to practice)?"* [15]

Here the Brilliant Prophet Scholar is a reference to Confucius, and the quotation is from the *Zhongyong* 中庸 chapter of the *Liji* 礼记. The reason why Cobo used this phrase was to propagate the basic idea of Christianity, for he wanted to convey the idea that there is only one unique and true religion.

We can also find many phrases from other ancient Chinese literature, such as *putian zhi xia, shuaitu zhi bin* 普天之下, 率土之滨 from the *Shijing* 诗经, *bu she zhouye* 不舍昼夜 from the *Lunyu* 论语, *gengtian zaojinq* 耕田凿井 from the *Jirang Ge* 击壤歌, *taiji* 太极 and *liangyi* 两仪 from the *Yijing* 易经, etc.

However, Juan Cobo sometimes employed commonly-used Chinese terms to express Western concepts, and at other times he created new combinations of Chinese characters in order to do so. One example is the phrase *heshang wang* 和尚王, which he introduced the word "bishop".

Another characteristic of the *Shilu*, as mentioned above, is that the author used science and logical reasoning to advance Christian doctrines. Although special emphasis was placed on the knowledge of geography, Juan Cobo also reported facts and results dealing with various natural sciences, including botany, zoology, geology, climatology, and medicine.

In conclusion, Juan Cobo's *Shilu* occupies a place of special significance in the history of science in China, primarily because it was the first book written in Chinese to introduce the Western religion, philosophy, and science together. Moreover, the approach Juan Cobo adopted in this book may have been a harbinger of those missionary tactics applied shortly thereafter by the Jesuits, namely their strategies known as "accommodation" and "knowing", which they pursued in hopes of convincing the Chinese to convert to the Christian religion.

---

[15] 光明先圣学者有曰："率性之谓道，修道之谓教。"性道无二致也，教其有二术乎哉？

BIBLIOGRAPHY

[1] Villarroel, F., ed., *Pien Cheng-Chiao Chen-Ch'uan Shi Lu* 辩正教真传实录, Spanish trans., Apologia de la Verdadera Religion, English trans., Testimony of the True Religion, Manila: UST Press, 1986.

[2] Jiménez, J. A. C., Spanish Friars in the Far East: Fray Juan Cobo and His Book *Shi Lu*, *Historia Scientiarum*, Vol.7-3, 1998, pp.181-198.

[3] Wang Chongmin 王重民, *Zhongguo Shanbenshu Tiyao* 中国善本书提要, Shanghai: Guji Press, 1983, p. 364.

[4] Pelliot, P., Notes sur quelques livres ou documents chinois conservés en Espagne, *T'oung Pao*, Vol. 26, 1928, esp. pp. 44-46.

[5] Fang Hao 方豪, *Zhongguo Tianzhujiaoshi Renwu Zhuan* 中国天主教史人物传, Vol.1, Beijing: Zhonghua Shuju, 1988, pp. 83-88.

THE CONTINUING INFLUENCE OF THE PORTUGUESE: THE CONTRIBUTION OF "A NEW INTERPRETATION OF WORLD GEOGRAPHY"

WANG Qianjin
Institute for the History of Natural Science, Chinese Academy of Sciences, Beijing

At the beginning of the 17th century, the Portuguese produced a large body of works designed to disseminate science, technology and culture in China, such as "An Explanation of the Celestial Sphere" by Emmanuel Diaz Junior, "Logic and Physics of the University of Coimbra" by François Furtado, and "True Doctrine of Music" by Thomas Pereira; Jean-François Cardoso worked in collaboration with others on preparing the "Map of A Complete View of Imperial Territory" of Kangxi, and Felix da Rocha and José de Espinha worked on drawing up the "Map of Qianlong Reign in Thirteen Rows". These works contributed to the development of science and technology in China, such matters frequently being studied by scholars.

However, the influence of the Portuguese in the development of science and technology in China was not limited to this. In the years after the Qianlong Reign, the Portuguese maintained their reputation in Chinese intellectual circles, but this has received little attention. The author of this paper seeks to demonstrate the continuing influence of the Portuguese in the development of science and technology in China through an analysis of "A New Interpretation of World Geography" by José Martins-Márquez.

José Martins-Márquez was a Portuguese interpreter in Macao. In 1858, he took part in the Tianjin negotiations as a member of the retinue of the French special envoy, Baron Jean Baptiste Louis Gros (1793-1870). In 1861, as an interpreter, he accompanied the Prussian ambassador, Count Friedrich Eulenburg (1815-1881), during his visit to China. These facts are evidence of his good command of and ability to speak Chinese.

In 1847, in Year 27 of the Daoguang Reign, José Martins-Márquez published his work “A New Interpretation of World Geography”, the “Haisan Xianguan”[1] edition in ten volumes, with around eight hundred thousand characters (see Figure 1). The first volume deals with geography, the globe, the rotation of the earth, the order of the five planets, the distance of the planets from the earth, new observations of the five planets, the five new planets and their distance from the earth, the phases of the moon, and solar and lunar eclipses, together with a new interpretation of eclipses as evidence of the spherical form of the earth. This volume also covers comets, stars and constellations, the Milky Way, the celestial and terrestrial equators, the system of 32 compass points, the seasons, latitude and longitude, parallels of long and short days, and the time lines of the earth.

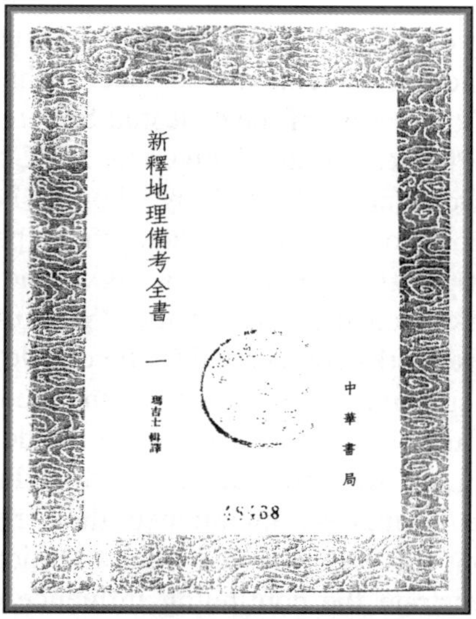

Fig. 1 – Title page of “A New Interpretation of World Geography” by José Martins-Márquez. [2]

The second volume deals with climate, clouds, winds, thunder and lightning, meteorites, rainbows, rings of light around planets, the rising of the sun and the moon, the aurora borealis, rain, snow, hail, dew, frost, fog, ice, tides, currents, springs, rivers, earthquakes, volcanoes, and so on. The third volume

[1] Divine Study of Seas and Mountains.

[2] Editor's note: the average quality of some of the illustrations of this paper are due to the originals provided by the author.

covers systems of state laws and political-economic geography, while the fourth to tenth volumes deal separately with the countries of Europe, Asia, Africa, America and Australia, describing their geographical location and areas, population, and geographical features such as mountains, rivers, lakes and seas, as well as their mineral resources, fauna, science and technology, religion, politics, history, customs and folklore. He also described the colonies of Britain, France, the Netherlands, and Belgium, among others (see Figure 2).

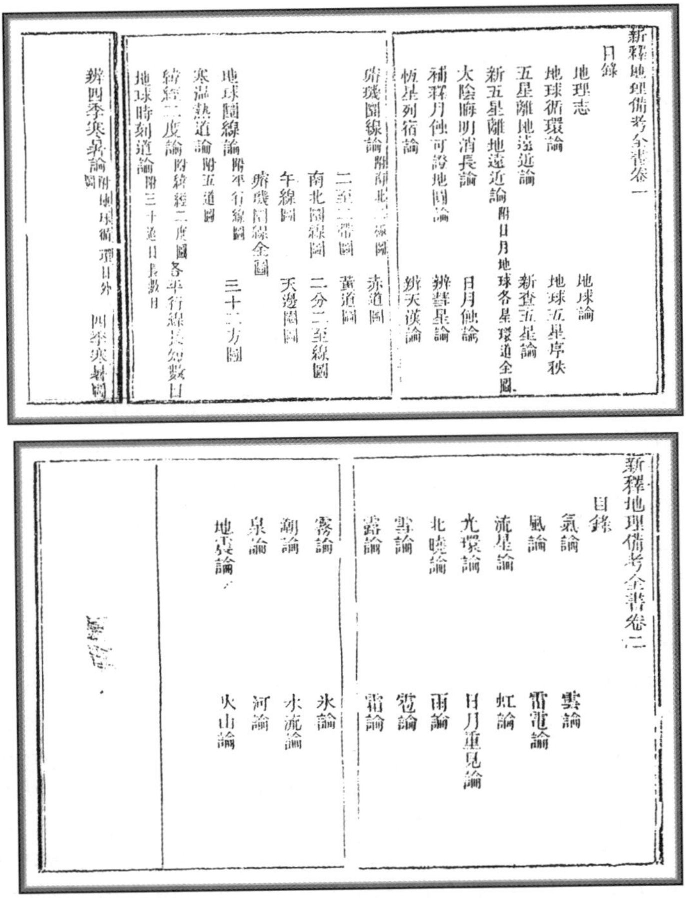

新釋地理備考全書卷一
目錄
地理志　地球論
地球循環論　地球五星序秩
五星離地遠近論　新查五星論
新五星離地遠近論 附日月地球各星環道全圖
太陰晦明消長論　日月蝕論
補釋月蝕可證地圓論　辨彗星論
恆星列宿論　辨天漢論
[illegible]圈線論 附[illegible]線圖　赤道圖
二至二帶圖　黃道圖
南北圈線圖　二分二至線圖
午線圖　天邊圈圖
[illegible]圈線全圖
地球圈線論 附平行線圖　三十二方圖
寒溫熱道論 附五道圖
結經二度論 附結經二度圖 各平行線長短數目
地球時刻道論 附三十道日長數目
辨四季寒暑論 附圖 附地球循環日外 四季寒暑圖

新釋地理備考全書卷二
目錄
氣論　雲論
風論　雷電論
流星論　虹論
光環論　日月重見論
北曉論　雨論
雪論　雹論
露論　霜論
霧論　冰論
潮論　水流論
泉論　河論
地震論　火山論

Fig. 2 – Contents pages of Volume I and II of José Martins-Márquez's work.

Prior to the work of José Martins-Marquez, there had been other geographical works translated by Europeans in China, such as "Complete Map of All Nations on Earth" by Matteo Ricci, 1552-1610, "Concise

Geography of the Whole World" by Jules Aleni, 1582-1649, "Treatise on Earthquakes" by Nicolas Longobardi, 1559-1645, "Treatise on the Material Composition of the Universe" by Alphonse Vagnoni, 1566-1640, "Explanation of the Earlier Map" by Ferdinand Verbiest, 1623-1688, "Mapa-Mundi" by Michel Benoist, 1715-1774. But what specifically characterised the work of José Martins-Márquez and what new knowledge did it include?

1. For the first time scientific rigour was applied to geographical information

In the chapter on geography in vol.1 of "A New Interpretation of World Geography", he states:

> *"Geography concerns the study of the earth, describing the geographical location of the countries of the world and giving the names of their rivers, mountains and seas, and can be divided into three categories:* cultura geográfica *(general geography),* característica geográfica *(specific geography) and* geografia geo-político *(geo-political geography). The first describes the South and North Poles, the northern and southern hemispheres, the Arctic and Antarctic Circles, the Tropics of Capricorn and Cancer, cold, temperate, subtropical and tropical zones, and latitudes and longitudes. Specific geography includes rivers, lakes, seas, mountains, valleys, plains, islands, straits and gulfs. Geo-political geography deals with countries, provinces, municipalities, districts, towns and parishes, political systems, number of inhabitants, titles of rulers, religion, and so on. These three main areas are essential for any study of the earth's geography. The origins of geography as a subject of study can be found in astronomy. It was only through astronomical studies that we discovered the topography of the earth, its degrees, and the differences between people from different parts of the world; thus, to gain greater knowledge, it is first necessary to imagine the earth as a planet that revolves around the sun at an enormous distance. Only then can we understand its relation to other planets and stars, and hence deduce its geographical characteristics."* (See Figure 3)

Thus, José Martins-Márquez divided geography into three areas: general geography, what today is known as mathematical geography; specific geography, now called physical geography; and geo-political geography, now known as human geography. In the various volumes of his work we can find detailed information on these three areas:

General geography: includes locations, borders and areas.

Specific geography: includes mountains, volcanoes, valleys, seas, gulfs, capes, straits, river basins, lagoons, rivers, lakes, springs, islands, peninsulas, roads, plains, deserts, climate, mineral resources, plants and animals (insects).

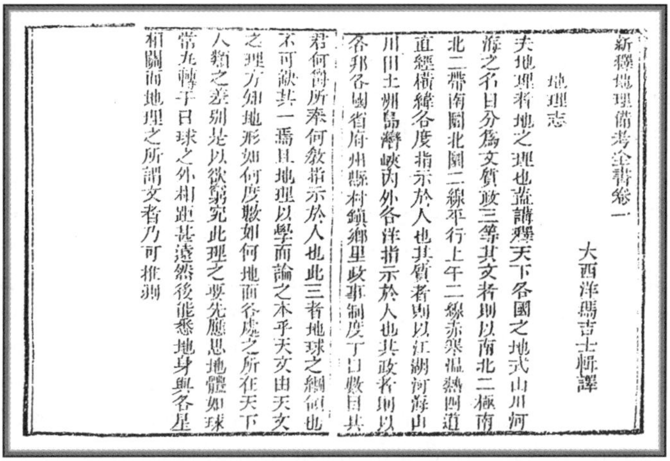

新釋地理備考全書卷一

大西洋瑪吉士輯譯

地理志

夫地理者地之理也蓋講釋天下各國之地式山川河海之名目分為文質政三等其文者則以南北二極南北二帶南圍北圍二線平行上午二線赤寒溫熱四道直經橫緯各度指示於人也其質者則以江湖河海山川田土洲島灣峽內外各洋指示於人也其政者則以各邦各國省府州縣村鎮鄉里政事制度丁口數目其君何爵所奉何教指示於人也此三者地球之綱領也不可缺其一焉且地理以學而論之本乎天文由天文之理方知地形如何度數如何地面各處之所在天下人類之差別是以欲窮究此理之要先應思地體如球常九轉于日球之外相距甚遠然後能悉地身與各星相關而地理之所謂文者乃可推測

Fig. 3 – The texts on Geography.

Geo-political geography: includes population, religion, regime (system of government), military organisation, cities and fortresses, art and culture, commerce, human characteristics and qualities, habits and customs, palaces and architecture, history, territory, states, ancient countries, capitals, provinces, islands and dependent territories.

Prior to the work of José Martins-Márquez, there were no such detailed descriptions in the Chinese language. For example, in the first volume of "Explanation of the Earlier Map" by Ferdinand Verbiest, it merely states:

> *" 'Explanation of the Earlier Map' describes the earth in general terms, in other words, its topography, earthquakes, mountain ranges, seas, sea quakes, rivers and peoples, together with their habits and customs, and products."*

This description is too general. Even after the publication of José Martins-Márquez's work, few treatises on geography surpassed it; for example, the chapter on "Geography" in vol. IV of "Outlines of Astronomy" by J.F.W. Herschel, published in 1849 by Alexander Wylie, in the translation by Li Shanlan, states:

> *"Geography is a part of astronomy, and in fact the most important one, since the earth is the point of reference from which to observe the sky; geographers study continents, islands, seas and oceans and the location of rivers and mountains, as well as geology, climate, products and peoples."*

The concepts of 'geographical studies' in the work of José Martins-Márquez were soon assimilated and developed by other authors. The Englishmen, Muirhead William, in his "Complete Geography", of which a Chinese translation was published in 1853, wrote:

> *"Geography is divided into three categories: general geography, specific geography and geo-political geography. The latter category has already been covered in the previous volume. We will now deal solely with geographical features, which concerns only the form and composition of the earth's rocks and their location, in which one studies the traces of birds, fish, animals and plants, and changes in continents and seas, soils and watercourses, mountain ranges, seas and oceans, unusual climate phenomena, peoples, animals and plants, which will be studied in detail as indicated below."*

Muirhead William himself acknowledged that his idea came from "A New Interpretation of World Geography". In the preface to his work, he states:

> *"Two books on geography have been published recently. The first is entitled "A New Interpretation of World Geography", written by José Martins-Márquez published by Haishanxinguan, and the other is "Ying Huan Zhi Lue", written by Xu Jishe published by the Prime Minister Office (Zong Li Ya Men) in Beijing. They are extremely clear and thorough, offering detailed evidence; those who discourse on geography seeks their ideas, terms and classifications there and those who write books on geography will often use them as reference works."*

From this, we know that the concepts of Muirhead William regarding general geography ans specific geography came from the work of José Martins-Márquez.

2. New knowledge in astronomy

2.1. The five new planets

The five new planets that José Martins-Márquez described were Uranus, Ceres, Pallas, Juno and Vesta, whose orbits lie between Mars and Jupiter (see Table 1 and Figure 4).

Western Name	Meaning of Chinese Name	Date of Discovery	Discoverer	Diameter *Li*[3]	Distance from Sun	Solar Year	Satellites
Uranus	God of the Sky	Year 46 of Qianlong Reign (1781)	F.W. Herschel Germany	122 120	662 000 in tens of thousands of *Li*	30 589 days 4 hours, two quarters, and 9 minutes	6
Ceres	Goddess of Cereals	Year 6 of Jiaqing Reign (1801)	G. Piazzi Italy	5420	95 522 in tens of thousands of *Li*	1680 days and 6 hours	
Pallas	Goddess of Wisdom	Year 7 of Jiaqing Reign (1802)	H.W.M. Olbers Germany	7000	95 892 in tens of thousands of *Li*	1680 days and 8 hours	
Juno	Queen of the Goddesses	Year 8 of Jiaqing Reign (1803)	C.L. Harding Germany	4750	91 278 in tens of thousands of *Li*	1588 days	
Vesta	Goddess of Fire	Year 12 of Jiaqing Reign (1807)	H.W.M. Olbers Germany		81 530 in tens of thousands of *Li*	1162 days and 2 hours	

Table 1

However, the work did not include Astrea, discovered by K.L. Hencke (1793-1866) in 1845, or Hebe, discovered in 1847; neither did it include Iris and Flora, discovered by the Englishman Hind in 1847.

[3] *Li* is a measure of distance equal to 500m.

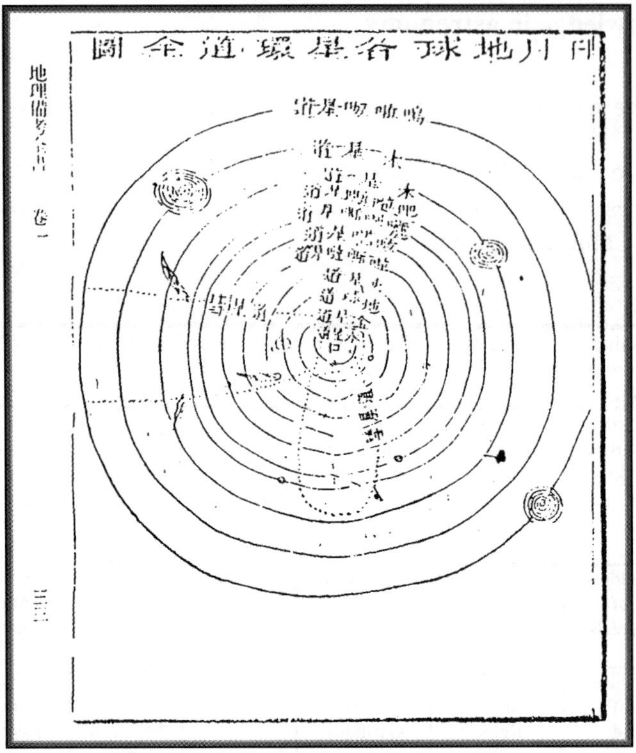

Fig. 4 – The Solar System.

2.2. The three 'returning' comets

In previous astronomical and geographical works, descriptions of comets consisted merely of defining them and the causes of their formation, but José Martins-Márquez went further. He described three comets, their cycles and the years they could be observed from earth. The cycle of the first comet is 76 years (this must be Halley's Comet), which was since in Year 10 of the Jiaqing Reign (1513), Year 35 of the Wanli Reign (1607), Year 21 of the Kangxi Reign (1682), Years 12-13 of the Qianlong Reign (1757-8), and Years 14-15 of the Daoguang Reign (1834-5). The second comet's cycle is 103 years, appearing in Year 11 of the Longqing Reign (this must be Year 5 of the Wanli Reign, 1577), Year 19 of the Kangxi Reign (1680), Year 48 of the Qianlong Reign (1783), and Year 66 of the Daoguang Reign (which must be Year 12 of the Guangxu Reign, 1886). For the third comet, the cycle was 34 years, being seen in Year 7 of the Kangxi Reign (1668), Year 41 of the Kangxi Reign (1702), Year 1 of the Qianlong Reign (1736), Year 35 of the Qianlong Reign (1770), Year 9 of the Jiaqing Reign (1804), Year 18 of the Daoguang Reign (1813), Year 52 of the Daoguang (this must be Year 11 of the Tongzhi Reign, 1872). It is through such concrete numbers that people can understand the wonders of astronomy and of nature.

3. New knowledge in meteorology and climatology

3.1. System of atmospheric circulation

In the second volume of his work, on "Wind", José Martins-Márquez wrote:

> *"The earth rotates from west to east, and the atmosphere above it moves from east to west, just as in the case of a ship travelling from west to east against the current, the water that passes along the hull of the ship will travel from east to west. Winds sometimes change for other reasons and they are therefore not all the same. For example, winds from the south and the north always blow in the same direction, with little variation; winds from the two regions between the equator and latitudes 30° south and north are of three types: the first are called prevailing winds, in the north they always come from the northeast, while in the south they always come from the southeast; near the equator, they move from the east towards the west. Even though the prevailing wind above the equator within the range 2° to 3° to the north and south is from the east, there are exceptions and some variations."*

In 1686, the Englishman E. Halley published his paper "Trade Winds and Monsoons", reasoning that the trade winds were related to the sun's heat at the equator. Two years later, in 1688, he produced a map showing the distribution of the trade winds and monsoons between latitudes 30°N and 30°S, based on data obtained on ocean winds; this is the oldest extant meteorological chart.

The results of Halley's studies on trade winds and monsoons were only improved upon in 1850 by Matthew Fontaine Maury (1806-1873), a North American naval officer. Maury collected data on ocean winds and currents, producing a model of atmospheric circulation in 1850. According to this model, the region around the equator between 30°N and 30°S is composed of a windless strip along the equator and a band of trade winds, which blow from the northeast in the northern hemisphere and from the southeast in the southern hemisphere. The zone between 30° and 60° in both hemispheres is a region with no tropical winds but with strong westerly winds, while the north and south polar regions are relatively calm (see Figure 5).

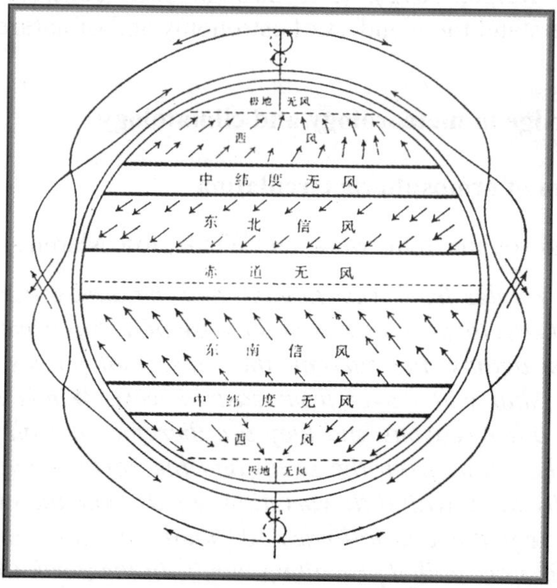

Fig. 5 – Diagram showing Matthew Fontaine Maury's system of atmospheric circulation.

3.2. Vertical change in atmospheric pressure

In "On the Atmosphere" in Volume II, it states:

"In higher regions, the air is rarefied and thinner, while at lower altitudes the air is denser and heavier. This is why people live in low-lying areas where they feel comfortable with this atmosphere; since the air is thinner in higher areas, people cannot survive. In the past, during the Jiaqing Reign, there were people from the West who used a balloon to go up into the sky; at a height of around 14-15 li above the surface of the earth, they began to experience a loss of hearing, accompanied by aching arms, while others felt innumerable types of sadness, difficult to describe since each case was different. If they went to greater heights, it was impossible to predict what might happen. Thus, people cannot go up too far from the surface of the earth."

This is a relatively clear description of vertical changes in atmospheric pressure, a phenomenon that was discovered in 1787 by the Swiss H.B. de Saussure (1740-1799). De Saussure founded the first high altitude meteorological observatory, on the lower reaches of the Entrèves glacier on Mont Blanc in the Alps. Through his observations he discovered that solar radiation increases, while atmospheric pressure falls, the greater the altitude.

3.3. Climatic characteristics of the different continents

While discussing the specific geography of the different continents in his work, José Martins-Márquez described the differences in climate; for example, in Volume IV on Europe, in the chapter "Climate":

> *"Most of Europe is in the northern temperate zone, with a twelfth of its area in the northern cold zone. Its climate can be divided into four types: the first, from 35° to 45°, is tropical; the second, from 45° to 55°, is temperate; the third, from 55° to 65°, is cold; and the fourth, from 65° to 72°, is extremely cold. However, even within these four categories — tropical, temperate, cold and extremely cold — there are always factors that influence the climate so that it is not so extreme. In countries on the Atlantic coasts, the cold increases from the south to the north, with ocean winds moderating temperatures. In the Mediterranean region, the temperature constantly switches between hot and cold. From the west towards the east, the cold gradually increases and the temperature changes constantly according to the winds. Eastern regions are slightly cooler than western ones. Areas bordering on Asia are extremely cold, with no great variation from south to north, and the heat is also intense. As regards the four seasons, the first, second and third types of climate have all four seasons, the only difference being that in some regions they arrive sooner and in others later. The fourth climate zone has only two seasons. The northern cold region is extremely cold — even in the months of June and July the waters freeze — and there are over three months when the sun does not set. There are even longer periods of darkness during which there is only the aurora borealis in the sky, and thus the winter is extremely long and cold since there is no sun or moon over a very long period of time. The summer is short but very hot, since the sun is constantly shining."*

In his work, the author divides Europe into four different climate zones according to latitude: 35–45°, 45–55°, 55–65° and 65–72°, describing the climatic characteristics of each zone, as well as the reasons for their formation which, apart from the difference in latitudes, is determined by their distance from the sea.

In Volume VII on Asia, he writes:

> *"The Climate: This continent has four different climates. From the North Pole to 62° it is extremely, unbearably cold, the sky is covered by mists and on the ground fungi grow everywhere. From 65° to 52°, the cold is less extreme but equally unbearable, with little vegetation and few cereals or fruits. From 52° to 35°, it is flatter, the climate is temperate with less extreme cold, and many cereals and fruits are grown. From 35° southwards, there are fewer cold days and more warm days, it is always spring or summer, with a great variety of cereals and fruits. In general terms, we can say of the five principal regions that the North is cold, the South is hot, the East is mild, the West is hot and dry, and the central region has very long, cold winters and very short, hot summers."*

Asia's climate is thus divided into four zones according to latitude: from 62° northwards, from 60° (or more accurately, from 62°) to 52°, from 52° to 35°, and from 35° to the southernmost point of Asia. In addition, he gives information according to the cardinal points: North is cold, South is hot, West is hot and dry, and East is mild, while the central regions are cold in winter and hot in summer.

Africa's climate is a special case. In Volume VIII on Africa, he states:

> *"The Climate: Of the five continents, Africa is the hottest since it is mainly on the equator with few areas in the temperate zone; areas near the sea are cooler, but in other places the heat is difficult to bear. There are great rivers which spread epidemics across the land; the extremes of hot and cold during the twelve hours of daylight is unbearable, even for the indigenous people, and foreigners frequently fall victim to disease. At the time of the change of seasons, in the midst of thunder, lightning, wind and rain, the heat diminishes slightly, but as soon as the sun sets, the heat returns to normal."*

The climate of Africa is the most scorching of the five continents, since "it is mainly on the equator with few areas in the temperate zone".

4. The causes of earthquakes

There are references to the causes of earthquakes in ancient Chinese texts, and with the introduction of Buddhism, the Indian view on why earthquakes occur also entered China, but all these are different from Western notions. In the section on earthquakes in Volume II of his work, José Martins-Márquez states:

> *"An earthquake is the burning of sulphur and other rocks in the heart of the earth. Indeed, it is the most dangerous catastrophe in the world, people have no means of predicting or avoiding it since there is no way of knowing where it will occur; it happens so fast that it is impossible to escape and even if occasionally there is time to escape, you do not know where to go. There is concrete proof that earthquakes are caused by the burning of sulphur and other rocks. A Westerner called Lemeili did an experiment, using iron filings, sulphur and water, a total of thirty-eight Jin[4]. After mixing them together well, he buried this compound in the earth and several days later the mixture exploded, the ground shook, and a raised mound appeared from which flames and smoke emerged. In Year 20 of the Qianlong Reign (1755), there was an enormous earthquake in the capital of Daxiyang[5] which caused great damage, rarely before seen in history. Many cracks appeared in the earth's surface, spewing out large quantities of black smoke and ash which smelled of sulphur, and the water boiling in the cracks also smelled of sulphur. As a general rule, any place that has hot springs or sulphur mines is prone to earthquakes. This is because ash, sulphur and other rocks in the heart of the earth have different characteristics, some are compatible with each other but others are not; for example, calcium and water are not compatible and when they meet, they react with each other. If the cavity in the earth is sufficiently large to disperse the energy generated by such reactions, then there is no earthquake; if the cavity is small, it is unable to disperse its energy, and will certainly cause movement in the surrounding area which will then disappear. If a cavity in the earth is burning, other cavities will be affected; as with gunpowder, igniting a small amount can cause an explosion*

[4] *Jin* is a measure of weight equal to 500g.

[5] Portugal

somewhere else. Explosions in the interior of the earth are also like this. So, when an earthquake occurs in one place, the same thing happens an instant later over an area of tens of li. If a cavity in the earth ignites, neighbouring areas vibrate, in the same way that a heavy weight hitting the middle of a wall will make the upper and lower parts on both sides shake. The harder the material, the greater the distance over which the vibration is transmitted. Earthquakes are also like this. Earthquakes are always accompanied by noise, the sound of steam caused by fires in the heart of the earth. Before large earthquakes, there are always signs: first, black clouds appear on the horizon which rise slowly until the sky is covered; second, there is great heat at the earth's surface, the air rises, expands and causes storms; and third, it is extremely hot, there is no wind, and all living beings become uneasy. Thus, all storms and earthquakes are always accompanied by omens in the form of abnormal phenomena; for example, the skies and the land can be very calm, with no sound, but the trees are moving although there is no wind — this is a sign of imminent danger. Another phenomenon that occurs when there are large-scale earthquakes is flooding of the sea and rivers in that region because the earthquake causes the waters to become abnormally agitated. As a rule, it is impossible to predict the extent and intensity of an earthquake. It is indeed a very special natural phenomenon."

It is not correct to define the cause of earthquakes as "*the burning of sulphur and other rocks in the heart of the earth*", but at least here the cause is sought in nature, the idea being that it is a natural phenomenon. This is far more scientific than the arguments put forward that earthquakes were warnings from God to humanity. Alphonse Vagnoni, in his "Treatise on the Material Composition of the Universe", described earthquakes as follows:

"The function of earthquakes is as a warning to people.

1) *The Creator sees that Man has forgotten himself and his origins and wishes to warn him so that he can mend his ways.*
2) *The cause of a place being completely turned around, when the top becomes the bottom and the bottom the top, when left becomes right and right becomes left, is subterranean cyclones.*

3) *The struggle between two forces causes two earthquakes, and that is why two mountains come together and destroy the houses in between.*
4) *During an earthquake, the earth opens up and devours a city and then immediately closes again over it.*
5) *At times, the earth does not open up but rises up like a mountain and does not go down again, forming new mountains and new islands at sea.*
6) *Great rifts open up at the bottom of the sea and these devour the waters of the sea as well as islands.*
7) *Great explosions of air at the bottom of seas and rivers cause the waters to form great floods as high as mountains which submerge people and property; the beds of rivers are laid bare, emptying their waters onto the banks or even returning to their source.*
8) *After an earthquake, new springs and streams appear, some hot springs become cold, other springs become hot, because the earth tremors open up new channels and close others, altering the circulation of underground waters and changing their characteristics.*
9) *Rivers of fire and volcanoes appear after an earthquake because the interior of the earth is hot and dry, and under great pressure. The combination of pressure and heat easily creates the conditions for fire. Therefore, when the fire bursts forth, it brings with it saltpetre, sulphur and black sand that have been produced by the heat.*
10) *The cracks produced by earthquakes emit fumes which pollute the air, causing epidemics; the air becomes turbid and the sky is covered with black clouds since the gases are extremely dense and turbid, and the sky does not clear for a long time. All the above is magic from Heaven; if it is not a warning to humanity, what purpose would it serve?"*

This is how Alphonse Vagnoni described the various phenomena associated with earthquakes, ending with a question.

The "*Dong Xi Yang Kao Mei Ri Tong Ji Zhuan*" (Research Journal of the Seas of the East and West), published in 1838, put forward the same view as Alphonse Vagnoni:

> *"The will of Heaven does not like to go to extremes, catastrophes and good fortune always follow one another, God rules over the world and no man, no being, can escape his control. If men respect and fear the Supreme Lord, obeying his orders, then a hundred joys and a thousand good fortunes will befall them. Catastrophes are caused by an accumulation of sins, evil acts that provoke the righteous wrath of the Lord. The Lord rules with great justice over the thousands of beings, He is all-powerful, all-knowing. If men are savage, disobedient, wilful, cruel and sinful, they will be punished with catastrophes, and the worst of these is the earthquake. In many mountains there is an accumulation of sulphur and combustible material that frequently burns, emitting smoke and flames, forming volcanoes. If the smoke and flames have no outlet from the mountain, the earth quakes in great surges of a thousand waves, the ground rises or falls, creating mountains and lakes in an instant, or making them suddenly disappear. Rocks disintegrate, hills collapse, islands rise up out of the sea following seaquakes, and rivers and seas flood the land, leaving rivers empty of their waters, all this in an instant. Cities and towns are destroyed, the land is devastated, people and animals die. This is why an earthquake is so catastrophic, so hard to describe. But all this is because men are evil and cruel, and they are being punished for their own sins. Think well on this: you should serve the Lord faithfully, so that these catastrophes will be transformed into good fortune."* (See Figure 6)

In the 17th and 18th centuries, some Western scientists believed the cause of earthquakes to be "fire" and others "water". In 1688, the English scientist Robert Hooke (1635-1703) wrote his "Discourse of Earthquakes", published in 1705. In this work, Hooke described the geological characteristics of earthquakes, volcanoes and the movement of continents. He believed that earthquakes were caused by subterranean fires and in turn that these subterranean fires took the form of volcanic eruptions. Earthquakes convert sea into land, and land into sea, plains into mountains and mountains into plains. The English geologist John Michell (1724-1793) published "Conjectures Concerning the Cause, and Observations upon the Phaenomenon of Earthquakes". He made a detailed study of the Lisbon earthquake of 1755 and reached the conclusion that the cause of the catastrophe lay in underground waters that had penetrated down to a certain depth and come into contact with volcanoes, thus creating a large quantity of gas and causing an explosion.

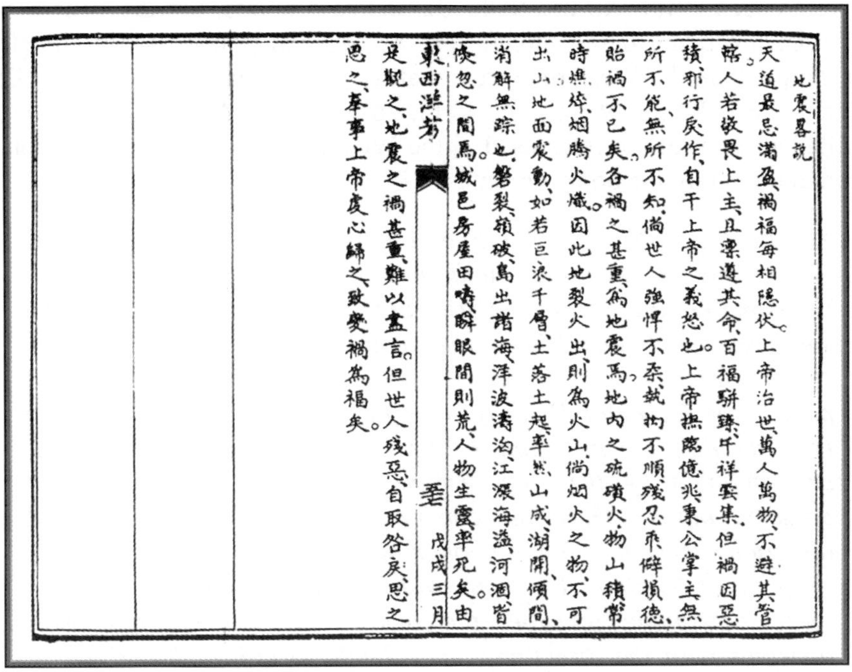

地震畧說

天道最忌滿盈、禍福每相隱伏。上帝治世、萬人萬物、不避其管轄。人若敬畏上主、且禀遵其命、百福駢臻、千祥雲集。但禍因惡積、邪行戾作、自干上帝之義怒也。上帝撫臨億兆、秉公掌主、無所不能、無所不知。倘世人強悍不柔、執拗不順、殘忍乖僻、損德、貽禍不已矣。各禍之甚重、爲地震焉。地內之硫磺火物山積、常時焆焠、烟騰火熾。因此地裂火出、則爲火山。倘烟火之物、不可出、地面震動、如若巨浪千層、土落土起、卒然山成、湖開、傾間消解無踪也。磐裂、嶺破、島出、諸海洋波濤洶、江溢海溢、河涸皆倏忽之間焉。城邑房屋田疇、瞬眼間則荒、人物生靈卒死矣。由是觀之、地震之禍甚重、難以盡言。但世人殘惡、自取咎戾、思之思之、奉事上帝虔心歸之、致變禍爲福矣。

東西洋考 戊戌三月

Fig. 6 – Texts concerning earthquakes from "Dong Xi Yang Kao Mei Ri Tongji Zhuan".

José Martins-Márquez assimilated these new theories from his contemporary John Michell's scientific study. Of course, neither water nor fire were the correct cause of earthquakes, but in his analysis, we can see his scientific approach.

José Martins-Márquez gave a detailed account of the four signs that predict the occurrence of large-scale earthquakes. His work contained observations and conclusions of scientific importance which went further than the various accounts in ancient China concerning the signs of the occurrence of an earthquake.

5. The 32 compass points

References to the 32 compass points began to appear during the Song Dynasty; in the "Bu Bitan" (Additional Commentaries) of Shen Kuo, there are references to how to locate the 32 compass points, but the system was different, not being based directly on the cardinal points. It was therefore less simple than the Western system that used a combination of the cardinal points - East, South, West and North. José Martins-Márquez presented a detailed account of the rules used in the West governing the combination of cardinal points. In his book, he states:

> *"In Portugal, France and England, the names of the compass points between the four main directions are obtained through the combination of two of these directions: southeast, northeast, southwest, northwest. This is the same as in the Celestial Dynasty (i.e. the Imperial Kingdom), but apart from these four compass points, the Celestial Dynasty does not have names to designate the remaining compass points. The 32 compass points of the West are obtained by the same means, through a combination of adjacent points. Thus, the combination of the two orthogonal directions form the four diagonal directions, the combination of the four orthogonal directions with the four diagonal directions form eight smaller divisions, the combinations of these eight small divisions with the directions immediately to either side give rise to sixteen adjacent directions, a total of 32 compass points."* (See Figure 7)

The use of the names of the four cardinal directions in combination to obtain names for 32 compass points makes them far easier to determine, memorise and read, while the traditional Chinese system of 32 compass points did not use the cardinal directions, but specific terms.

Conclusion

"A New Interpretation of World Geography" by José Martins-Márquez is a work which links the past and the future in the sense that in many respects it went beyond previously translated geographical works, and the names and technical and geographical terms used by him were adopted in later geographical works.

The publication of this work by José Martins-Márquez also illustrates the continuing influence of the Portuguese during the first half of the 19th century in the introduction of Western knowledge into China, making a positive contribution to the development of Chinese science and technology.

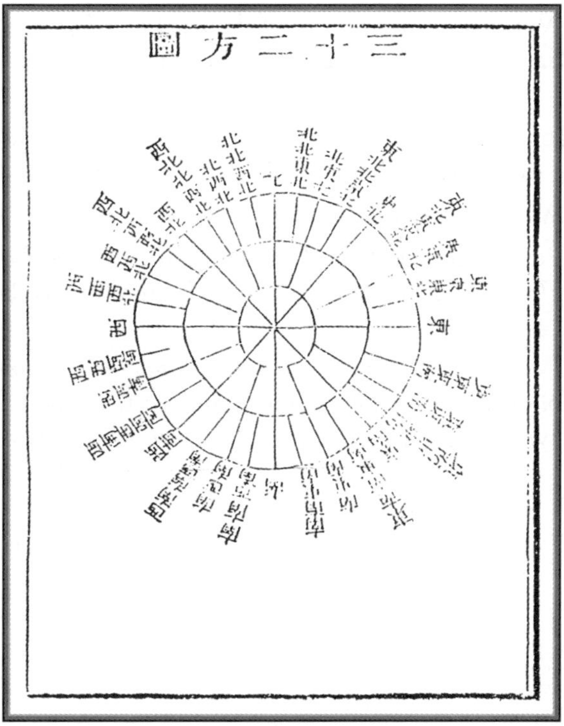

Fig. 7 – Diagram showing the 32 compass points.

BIBLIOGRAPHY

[1] José Martins-Márquez: *A New Interpretation of World Geography*, Shanghai: The Commercial Press, Ltd (Copy from first edition of the Haishan Xianguan collections).

[2] *Dong Xi Yang Kao Mei Ri Tong Ji Zhuan* (Research Journal of the Seas of the East and West), Beijing: (Zhonghua Shuju), photolithographic edition, 1997.

[3] Ferdinand Verbiest: *Explanation of the Earlier Map*, first edition of the Haishan Xianguan collections, Shanghai, The Commercial Press, Ltd, 1937.

[4] John Herschel: (translation by Alexander Wylie and Li Shanlan) *Outlines of Astronomy*, Shanghai, The Commercial Press, Ltd. 1930.

[5] Muirhead William: *Complete Geography*, 1853 edition.

[6] Preston E. James and Geoffrey J. Martin: *All Possible Worlds: A History of Geographical Ideas* (second edition), John Wiley & Sons Inc., N.Y., 1981.

[7] Alphonse Vagnoni: *Treatise on the Material Composition of the Universe*, photolithographic edition based on the woodcut edition from the Ming Dynasty, *First Collection of Astronomy*, Taiwan.

English translation by Paul Covill from Portuguese version of Chinese text.

TEACHERS OF MATHEMATICS IN CHINA: THE JESUITS AND THEIR TEXTBOOKS (1580-1723)

Catherine JAMI
CNRS, Paris

The sciences and the construction of a Jesuit identity in China

The transmission of some elements of the European sciences to China, which started in the late sixteenth century, was perhaps the largest scale and most systematic enterprise of the kind in Jesuit missions of the time, both in Asia and America.[1] This transmission can be seen as one aspect of the Jesuits' "accommodation" to Chinese culture. In China (just as in Europe) they sought the support of the elite, striving to reach the centre of power: they hoped that, like the Roman Empire, the Middle Kingdom would be evangelised from the top down if only they could convert its sovereign. Their transmission of the European sciences began, and was gradually turned into a policy, after Matteo Ricci (1552-1610), the founder of the mission, noticed the interest of literati who visited him for the world map hung on the wall of his room.

In terms of the construction of their identity in China, the Jesuits needed to differentiate themselves from the Portuguese, settled in Macao since 1511. The latter were regarded as Barbarians who did not respect the laws of the Empire. But the Jesuits also needed to distinguish themselves from those who turned out to be their main opponents and competitors, namely the Buddhist priests and monks. The well-known story of Ricci, first dressed as a Buddhist priest, then discarding this attire in favour of a literatus gown, reflected his change of strategy: from then on he strove to appear as a

[1] In what follows, the phrases "the European sciences" and "the Chinese sciences" refer to the categories used by Valignano and Ricci (see quotations infra).

scholar rather than as the member of a clergy and to move in literati circles.[2] Ricci's success in this respect can be assessed from the impression he made on the radical philosopher Li Zhi (1527-1602):[3]

> *"There is none of our books that he hasn't read. He has asked an elderly man to set for him the sounds and meanings* (of Chinese characters)*. He has asked someone expert in the philosophy of the Four Books to explain to him their general meaning. He has asked someone versed in the Commentaries of the Six Classics to provide him with the necessary explanations. He is now perfectly capable of speaking our language, of writing our characters, and of following our customs. He is a most remarkable man. [...] All are inferior to him. But I don't quite know what he is doing here. I have already met him three times, and I still don't know what he is doing here. I think that if he wanted to substitute his own teachings to those of the Duke of Zhou and Confucius, that would be too stupid. So it must not be that."* [4]

Ricci's culture and behaviour made him a legitimate member of elite circles; that he might be aiming at replacing Confucian orthodoxy by Christianity was beyond the imagination of even Li Zhi, who was highly critical of the Neo-confucian tradition. We shall see below how the Jesuits and some Chinese converts articulated Christianity's compatibility with Confucian teachings, and the role played by the sciences in that articulation.

Li Zhi's perplexity as to the purpose of Ricci's presence in China points to the necessity for the missionaries to find themselves a niche in literati circles. Ricci's construction of himself as a teacher of the European sciences provided such a niche. And Jesuit assessments of the Chinese sciences had paved the way for him to assume the role of a teacher. Thus, Alexandro Valignano (1538-1606), the Society's visitor of the East Indian and Japan province, who commissioned Ricci enter China, wrote:

> *"Although their sciences are not so perfect as ours, they appear to have reached the degree achieved by the ancient philosophers, before Aristotle introduced into them method and before they had been clarified in the light of the Christian doctrine."* [5]

---

[2] Ricci and Trigault 1978: 337-338. Standaert: 544.

[3] On Li Zhi, see Billeter 1979.

[4] Quoted by Gernet 1982: 30.

[5] Valignano, "De Chinensium admirandis", quoted by Bernard 1935: 39.

Valignano located the Chinese sciences on a historical path of progress, a path which, needless to say, led to the situation of Christendom in his own time. In other words, the Jesuits' arrival was timely: the Chinese were ready to study, understand and accept the late Renaissance Christian world-view, which included the Aristotelian philosophical tradition. The latter topic being part of the education given in Jesuit colleges, Ricci and his confrères were indeed in an excellent position to teach it in China. Ricci's characterisation of the Chinese sciences, not surprisingly, also pointed out to what they left to be desired:

> *"They have acquired quite a good mastery not only of moral philosophy, but also of astrology* (i.e. astronomy) *and of several mathematical disciplines. However, in the past they have been better versed in arithmetic than in geometry; but they have acquired it and dealt with it in a confused way."* [6]

Whereas Valignano referred to the late Renaissance curriculum in general terms, Ricci, a former student of Clavius (1538-1612) at the Roman College, identified the mathematical sciences as a field in which the Chinese both needed and were able to learn. The Society of Jesus had constructed the largest network of education in Europe. Teaching was part of the Jesuit training, so that the China missionaries had had some teaching experience in one or several of the Society's colleges in Europe, or on their way to China. Once there, they merely reinterpreted one of their European roles. But they were by no means in a position to set up colleges similar to the European ones or to those of Goa and Macao, in a country which had its own education system. For they were obviously incapable of teaching the Classics, which were the core of education in China. Instead they used mathematics, and related topics such as cartography, to construct master-disciple relationships with a number of literati.[7]

In what follows, I would like to discuss some aspects of these relationships, in the cases of the two Jesuit missionaries best known for having taught mathematics in the late Ming period (that is, up to 1644): Ricci (Chinese name Li Madou), and Giulio Aleni (1582-1649, Chinese name Ai Rulüe). Whereas both of them had studied at the Roman College, they worked in different geographical and social contexts. After them, I will consider the case of Nikolaus Smogulecki (1610-1656), and then turn to the

[6] Ricci and Trigault 1978: 95.

[7] On this issue see also Baldini 2000: 89.

textbooks written for the Kangxi Emperor (r. 1662-1722). Before closing, I will briefly evoke some controversies aroused by the teaching of the mathematical sciences within the Society of Jesus.

The first student

The first person to be mentioned in Ricci's Journals as having been his student in mathematics was Qu Rukui (1549-1611).[8] It was in 1589, when Ricci was in Shaozhou, that Qu Rukui first approached him:

> *"Then one day, in solemn splendour following the Chinese custom, carrying precious gifts, he came to Father Matteo, to elect him as his master according to the way of the country. The use is to proceed it as we have said in Book One when dealing with customs and ceremonies. The next day, having invited him to a feast, he received him in his cell, and the Father could refuse nothing of this, not even the gift, most of which was silk cloth; but he rewarded him amply by other things from Europe, so that he would not think that the Fathers were attracted to teach men by the love of riches. His main design, although from the beginning he held it very secret, tended towards alchemy. For this opinion by which it was thought that our fathers made silver had not at all been removed from men's mind. But that one, being disabused of this mistake by the daily contact with our fathers, applied his mind to more elevated and more serious sciences. He started with arithmetic, which in method and ease by far surpasses the Chinese one. For the latter all in all consists in a certain wooden instrument to which round beads, strung on copper wire, are changed here and there, to mark numbers. Although in fact it is sure, it is easily subject to misuse, and reduces a broad science to very little. He then heard Christoph Clavius' sphere and Euclid's elements,[9] only what is contained in Book I; towards the end he learnt to paint almost any kind of figures of dials to mark the hours. He also acquired knowledge of the heights of things through the rules and measures of geometry. And being, as I said, a man of wit and well versed in writing, he reduced all this into commentaries in a very neat and elegant language, which he later showed to*

[8] On Qu Rukui (Taisu), see Huang Yilong 1994.

[9] Ricci was using the Latin edition by his master Clavius (Rome, 1574).

> *mandarins. One would hardly believe what reputation this earned to him and to our fathers, from whom he acknowledged having learned it all. For all that he had been taught delighted the Chinese, so that it seemed that he himself could never have enough of learning. For he repeated day and night what he had heard, or adorned the beginnings with figures so beautiful that they were by no means inferior to those of our Europe. He also made several instruments, spheres, astrolabes, dials, magnet boxes, mathematical, and other similar instruments very elegantly and artistically set up. And the material too was by no means inferior to the work for, not content with wood and copper, he made several of silver."* [10]

Qu Rukui became one of the Jesuits' advisors and sympathisers, and eventually converted in 1605. But he was first attracted to them for purposes that were neither religious, nor scholarly. The metamorphosis he underwent was spectacular: having approached Ricci in the hope of acquiring silver, [11] he ended up contributing silver. This noble material enhanced the status of the instruments, which thus superseded the abacus, the latter being made of plain wood and copper. When Qu first sought Ricci's acquaintance, the latter had just been expelled from Zhaoqing, and the former was an outcast.[12] By heralding Ricci as a great master, he may also have sought to reconstruct his own status in Chinese society.

In Ricci's mind a prerequisite to Qu Rukui's first study, that of arithmetic, was the discarding of the abacus, then the main if not only calculation device in use in China, in favour of written calculation. In this particular case, Ricci's confidence of the superiority of the latter only leads one to suspect that he did not master the use of the abacus.[13] On the other hand, written calculation, which never supplanted the abacus in China, did provide literati with a distinctive technique. This technique enabled them to reconstruct "scholarly mathematics" — as opposed to "popular mathematics".[14]

---

[10] Ricci and Trigault 1978: 307-308.

[11] Engelfriet 1998: 58 analyses Qu's interest in alchemy in the context of late Ming Daoism.

[12] He had been expelled from his family following an accusation that he was having an affair with his sister-in-law. Huang Yilong 1994.

[13] On this issue, see Jami 1992.

[14] The Ming dynasty is usually regarded as a time of decline for mathematics, when the major works of the Song-Yuan period, especially those which expounded *tianyuan* algebra, were lost or no longer understood. In contrast, "popular mathematics", based on abacus calculation, underwent a significant development.

Regarding geometry and cosmology, Qu Rukui's study consisted in listening and in repeating until he eventually became able to produce written texts. To the best of my knowledge none of them have been preserved, so that it is impossible to assess his teacher's praise of their style and of the figures included in them. In any case the evidence for the success of this mathematical education was not the student's capacity to do calculations or geometrical reasonings, but his capacity to produce texts — which the teacher would have been unable to write on his own — and instruments. These were tangible results that served to advertise for the master and for his learning.[15] Qu Rukui introduced several scholars to the Jesuits' mathematics. And according to Ricci, after ten years, "from disciple, he had turned into a small master".[16]

The first doctor-student

In 1607, almost twenty years after Qu Rukui first sought to become his disciple, Ricci, together with another student, published the translation of the six first books of Euclid's *Elements*.[17] This was neither the first translation of Euclid, nor the first mathematical book published by the Jesuits. Qu Rukui is said to have translated the first book of the *Elements*. Ricci also mentioned that he had some mathematical treatises in Chinese printed while he was in Nanjing between 1598 and 1601.[18] But the 1607 translation is the earliest work still extant.

The social status of the student in question, Xu Guangqi (1562-1633), as well as the circumstances in which the teaching took place, seem to epitomise Ricci's progression towards the centre of power. The translation was done in Beijing, between 1604 and 1607, that is, during the three years of Xu's appointment at the Hanlin Academy following his success in the metropolitan examination, which opened the way to a career of higher official. Once a *jinshi* (a "Doctor", in the Jesuits' words) Xu, a convert since 1603, was in a position to protect and help the Jesuit missionaries at quite a different level than Qu Rukui had been able to. Thus in his late years, Xu would supervise the calendar reform on which some Jesuits worked at the

---

[15] Engelfriet 1998: 61.

[16] Ricci & Trigault 1978: 411.

[17] On this translation see Engelfriet 1998.

[18] Ricci & Trigault 1978: 412.

Astronomical Bureau.[19] Indeed he provided an excellent stepping stone for Jesuits mathematics. His fame may explain the neglect, in later accounts, of the previous mathematical works mentioned above: Xu Guangqi and Euclid's *Elements* seem to have enhanced each other's prestige in received historiography.[20]

Following the translation technique for Buddhist texts a millennium before, and for many other works published in China by the Jesuits, the translation in question, entitled *Jihe yuanben*, was "taught orally" (*koushou*) by Ricci and "received in writing" (*bishou*) by Xu Guangqi. This is how Ricci described this collaboration:

> "Ciù *Paul (Xu Guangqi), having given every day one hour to hear Father Matteo's lessons, progressed so much by the latter's care and diligence that he laid down in very beautiful Chinese language what he had hitherto been able to understand. Within one year he thus put* (wrote) *the first six books of the* Elements *clearly and neatly in Chinese. For this language is not short of words to explain well the meaning of all our sciences. He would well have continued with the other books of Euclid, but these seemed to suffice Father Matteo for his purpose."* [21]

As in Qu Rukui's case, the completion of a text in literary Chinese bore witness to Ricci's success as Xu's geometry teacher. This is in contrast with Li Zhi's account quoted above, in which Ricci's teacher in the Classics, although referred to respectfully — an elderly man —, remains anonymous, the merit seeming to lay mainly on the student's side.

Other mathematical works were published after the *Jihe yuanben*. Ricci "taught orally" not only Clavius' edition of Euclid's *Elements*, but several of the Clavius' textbooks, some of which the latter must have used when he had taught the former at the Roman College.[22] It is hardly surprising, then, that Ricci depicted both Qu Rukui and Xu Guangqi as listeners who dutifully wrote down what they had heard, and the aim of his own teaching as the production of textbooks that could attract more students. This was part of Ricci's apologia of his own proselytisation strategy, which underlay his

---

[19] Hashimoto 1988.

[20] An exception is Engelfriet 1998:

[21] Ricci and Trigault 1978: 570. According to the curriculum of the Roman College at that time, the first six volumes of the *Elements* were taught. Ricci was one of Clavius' early students at the College.

[22] Standaert 2001: 739.

whole account of the "Christian expedition to China". But it may also have reflected criteria of mathematics teachers for assessing their students in Jesuit colleges.

Xu Guangqi's social status and career enabled him to play a prominent role in the construction of "heavenly learning" (*tianxue*) — as Jesuit teachings came to be known — in terms of the Neo-confucian scholarly tradition. In his preface to the *Jihe yuanben*, he described the Jesuits' teachings as follows:

> *"As to the Master's (*xiansheng, *i.e.* Ricci*) learning (*xue*), it is three-fold: the greater part is self-cultivation and serving heaven (*xiushen shitian*), the lesser is the investigation of things and probing of principles (*gewu qiongli*). One distinct doctrine (*duan*) of the principles of things (*wuli*) is images and numbers (*xiangshu*), each point of which is essentially solid and authoritative; moreover it can make one certain beyond doubt."* [23]

While using Neo-confucian categories, Xu indeed appropriated the hierarchy between moral and religion on the one hand, and worldly learning on the other hand: again, both pertain to learning, a Confucian pursuit *par excellence*. Whereas Ricci's civility and erudition had been enough to differentiate him from the Macao Portuguese, it took the orthodox phrasing of a *jinshi* to make the *whole* of his teachings part of "learning". In this preface and in many other writings, Xu proposed an answer to Li Zhi's perplexity concerning what Ricci had come to China for. Far from being a threat to orthodoxy, Christianity could "do away with Buddhism and complement Confucianism" (*qufo buru*).

Not only were Ricci's teachings orthodox, but his credibility as a teacher also stemmed from the fact that he was the heir of a long tradition of scholarship, a legacy which he was in turn transmitting in China:

> *"Since his youth, Master Li has devoted the leisure that the study of moral philosophy has left him to this study. And it can really be said that in his country the master transmits and the disciples apply themselves. His master, Mr Ding,*[24] *is moreover a peerless scholar, therefore he is utterly well versed in these matters."* [25]

---

[23] Wang Zhongmin 1984: 75.

[24] A translation of Clavius' name.

[25] Wang Zhongmin 1984: 75.

Not only the contents of Ricci's teachings, but also the context in which he had received it, matched the received pattern of learning. In the same preface, Xu also pointed out that mathematics had been a concern of sovereigns and sages of the Chinese antiquity, and that it was an indispensable tool of *jingshi* (lit: "ordering the world"). In that sense, mathematics responded to the demand of a number of late Ming scholars and officials, whose concern with literati's role in the society had lead them to study technical subjects. Only once it had been properly translated into their language and categories by Ricci's "students" could "Heavenly learning" provide a response to such a concern.

Master Ai's teachings

After the *Jihe yuanben*, only one other work on geometry was published before the fall of the Ming dynasty. The *Jihe yaofa* (1631) had been "taught orally" by Giulio Aleni and "received in writing" by Qu Shigu (1593-?), who was Qu Rukui's son, in 1623-24. It was not a translation, but a compilation of some geometrical constructions of *Jihe yuanben.* [26] This intention is implied by the work's title: "*yaofa*", literally "essential methods" (*fa* was the Chinese rendering of *praxis*). A preface by Zheng Hongyou, dated 1631, states that the work was written as a summary of the *Jihe yuanben*, aiming at making the latter's meaning clear and easy. This suggests that Aleni and a number of Chinese readers of the *Jihe yuanben* had a shared understanding of what the essentials of geometry were: not the proofs, but the constructions. Zheng Hongyou also explains that when he met Aleni and saw the *Jihe yaofa* in 1631, he decided to have the work printed.[27] This brings us to the context in which the *Jihe yaofa* was written: unlike Ricci, Aleni worked outside the capital (either in Hangzhou or in Changshu), with a member of the local gentry whose patronage was not sufficient to have the work printed. The presentation of geometry proposed in the work, on the other hand, seems to have been adapted to such local gentry, rather than to high officials or specialised scholars.

Whereas I do not know of any Western sources recounting Aleni's activities as a teacher, the *Kouduo richao* ("Diary of Oral Admonitions") compiled by Li Jiubiao (?-1646?), a Fujian convert, reports dialogues

[26] Jami 1997.

[27] The *Jihe yaofa* was eventually included in *Xiyang xinfa lishu* ("Calendar Compendium According to the New Method from the West").

between Aleni and some members of the local gentry, which took place between 1630 and 1640. Although the work mainly deals with religious and moral issues, a number of the dialogues transcribed in it concern astronomy and other sciences.[28] They give us an idea of what sort of knowledge a former student of Clavius could convey in such circumstances. As an example, this is how Aleni explained the variation of the duration of day and night according to latitude:[29]

> *"On the evening of the 13th day (of the 3rd month, Chongzhen 3, that is, 1631), Wang Zijian and I, while waiting for the Master (*xiansheng*), happened to be discussing "being earnest in study and sparing of time," and theories of the length of day and night. Master Ai said:*
>
> > *"In Master Lu's Northern Europe,[30] at times the days are extremely short and the nights extremely long."*
>
> *I exclaimed in bewilderment. The Master said:*
>
> > *"Don't be so amazed: moreover there are lands at the North Pole where half of the year is day and half of the year is night."*
>
> *I exclaimed even more at this. The Master said, laughing:*
>
> > *"Long ago, when Mr Xu* (Guangqi) *heard this theory, he deeply sought without catching it. I took an astrolabe and showed it to him, only then did he begin to understand."*
>
> *I asked for a sketchy explanation. The Master said:*
>
> > *"These lands are right under the North Pole. From the spring equinox to the autumn equinox, the Sun revolves on the upper side of the Earth and fully shines down on the lands below the North Pole. Thus during this half-year it is always daylight and there is no night. From the autumn equinox to the spring equinox, the Sun revolves on the lower side of the Earth and does not reach the lands below the North Pole. Thus during this half-year it is always night and there is no day. If one were at the South*

---

[28] Zürcher 1997, esp. 606.

[29] Li Jiubiao, *juan* 1: 5a-5b. I am grateful to Linda de Lange who provided copies of the relevant passages of *Kouduo richao* and shared her work with me. The responsibility for the translation below, however, is solely mine.

[30] This is Andre Rudomina (1594-1632, Chinese name Lu Ande), who worked in Fujian with Aleni.

> *Pole, it would be the opposite. Now in the records of your Middle Country, it is written that "If you start steaming a sheep's shoulder bone when the sun comes out, the sun will have set before it is done." By and large, this is all about lands [close to] the North Pole."*

*Hearing this dazed me out of my wits. I withdrew and reflected upon what was drawn on the map, and somewhat understood what this meant. Now opinionated scholars say that what they have not seen with their own eyes does not exist; how would they know that if one fathoms things according to the principle (*li*), there are some that can be examined solidly and indisputably like this?"*

This is not a class on astronomy, but an apparently informal discussion. No textbook is involved here: Master Ai's teachings were oral, like that of Confucius, the model teacher. The conversation starts between two disciples, and the Master intervenes to deliver a bewildering but authoritative statement concerning the issue. He does provide further explanation when asked for it; however the meaning of his words only dawn upon the disciple gradually. At first the latter is dazed, but does not express disbelief: it seems that the Master's authority is well established, so that his assertions are not questioned. In this case however, Aleni invokes the precedent of Xu Guangqi, and to the astronomical instrument he showed him, to give weight to his assertion. Here the instrument serves as a mere rhetorical device, as Aleni neither produces it nor provides the corresponding technical explanation for Li and Wang's benefit. The fact that a scholar of high repute like Xu had understood, and accepted, Aleni's ideas on the variability of the lengths of day and night, seems to restore Li Jiubiao's faculties of speech and reasoning. Having recovered from his amazement, he humbly asks for "a simple explanation". In the end it is by looking at a two-dimensional map, not at a terrestrial globe, that Li Jiubiao "somewhat understands". His readers are in turn provided with an elementary description, which refers neither to technical astronomy, nor to any cosmological model.

Astronomy and mathematics sometimes came up in the civil examination questions. Therefore they had to be part of the literati's general knowledge, at least in a non-technical way.[31] The Jesuits' teaching of technical subjects, which entailed the production of textbooks, responded to the demand of a small number of scholars who engaged in such studies.

---

[31] Elman 1998.

On the other hand the dialogue quoted above, and more generally, Ricci's and Aleni's conversations with literati on "natural philosophy", show that they both interfered aptly in late Ming literati's concerns. It took a collection of specialised works such as the *Tianxue chuhan* (*First Collectanea of Heavenly Learning*, compiled in 1626 by Li Zhizao (1565-1630), another famous doctor-student of Ricci's) to investigate all the fields belonging to heavenly learning. In contradistinction, these fields could be displayed in a single social conversation, or in a single work like the *Kouduo richao*. This was how a wider audience was reached, this was what made the non-technical conversations recorded in the latter work an essential part of the construction of the Jesuit identity in China.

The teachings of a "Sincere gentleman"

In 1629, several Jesuits started to work on a calendar reform under Xu Guangqi's supervision.[32] From then on, their effort was mainly oriented towards imperial institutions, and only few cases of Chinese scholars studying with them are known. The most notable case concerns Nikolaus Smogulecki (1610-1656). The context was that of the dynastic transition: after residing in Fujian, where he witnessed the ongoing Sino-Manchu war, he resided in Nanjing from 1651 to 1653. It was during these two years that he worked on some mathematical and astronomical treatises, together with Xue Fengzuo (1600-1680). Unlike Qu Rukui, Xu Guangqi, Qu Shigu, and Li Jiubiao, Xue Fengzuo and other literati known to have discussed scientific topics with Smogulecki in Nanjing never converted. Possibly as a result of this, the *Siku quanshu zongmu tiyao* (1782), which gave an "imperially authorised" evaluation of all works, assessed him quite positively:

> *"He liked discussing mathematics (*suanshu*) with scholars and did not enrol them into the Society of Jesus. In his teachings, he was styled 'Sincere Gentleman' (*Dushi junzi*)."* [33]

Fang Hao (1910-1980), the famous historian of Christianity in China, has tried to refute this assessment, which was no doubt intended as a compliment by the compilers of *Siku quanshu zongmu tiyao*. What interests us here is the perception by later generations of Chinese scholars that

[32] See Hashimoto 1998.

[33] *Siku quanshu zongmu*, vol 1: 899-900.

mathematics was a gentlemanly pursuit, as opposed to the promotion of Christianity. In other words, a dividing line was drawn among "heavenly learning". Mathematics on the one hand, and what Xu Guangqi had called "self-cultivation and serving heaven" on the other hand, fell on opposite sides of this line. The assessment quoted above may also have reflected a dividing line within Smogulecki's activities. It may be that he was sharing his time between scholars, with whom he did not promote Christianity, and commoners or lower gentry, to whom he devoted his missionary activities.[34] Such a division of tasks had to be adopted when the Jesuits started working as imperial astronomers, cartographers, teachers, and craftsmen, for the new dynasty.

Smogulecki's teachings have aroused other historiographical controversies. These regard his attitude towards Copernicanism,[35] and his translation of some astrological treatises. Historians of science have argued that he introduced the heliocentric theory, whereas missionary-historians have long denied that he had ever introduced European astrology, unconvincingly in both cases. [36]

Smogulecki's teachings in astronomy were published under the title *Tianbu zhenyuan* ("True Source of the Pacing of Heavens"), as part of Xue Fengzuo's *Tianxue huitong* ("Integration of Calendar Studies", 1664). The first page mentions that the work was "translated orally" (*kouyi*) by Smogulecki and "compiled" (*bianji*) by Xue Fengzuo. This suggests a perception of their working relationship different from those between Ricci and Aleni on the one hand and their students on the other hand. Besides the fact that Xue Fengzuo never converted, one reason for this difference may have been that he was already well-versed in astronomy when he met Smogulecki. Having studied astronomy with Wei Wenkui, known as an opponent to the Jesuits' calendar reform, he adopted the Western methods which Smogulecki had introduced to him. Other scholars who met the latter in Nanjing, such as the mathematician Fang Zhongtong (1633-1698), were well versed in both Chinese and Western learning, which they strove to bring together.[37] To them, the Jesuits could no longer appear as the sole source of valid knowledge in mathematics and astronomy.

[34] The Chinese works attributed to him all have to do with astronomy, mathematics and cartography.
[35] See Shi Yunli 2000.
[36] The latter issue is now settled by the identification of his European source for astrology. Standaert forthcoming.
[37] Engelfriet 1998: 362-366.

From Jesuit treatises to the Imperial compendium

Since the advent of the Manchu dynasty in 1644, and except for a few years in the 1660s, one or several Jesuits were serving as officials at the Astronomical Bureau (*Qintianjian*).[38] Ferdinand Verbiest (1623-1688) who was Administrator of the Calendar (*zhili lifa*) from 1669 to 1688, gave impressive figures as to the number of students he had (these were the Bureau's "trainees" (*tianwen sheng*)): between 100 and 300 depending on the years.[39] According to recent research, a significant number of the Bureau's staff were Christians.[40] In their case the correlation between following the Jesuits' "greater learning" and "lesser learning" still held. But none of these disciples is known to have left writings of his own. Moreover, they formed a small profession, which had little or no connection with the scholarly gentry then active in the pursuit of the sciences.

On the other hand, Verbiest found his most powerful and prestigious student in the person of the Kangxi Emperor. From 1670 to 1674, the former taught the latter geometry; a translation of *Jihe yuanben* into Manchu was undertaken.[41] In the 1690s, Antoine Thomas (1644-1709) and Tome Pereira (1645-1708) — who jointly took up Verbiest's task at the Astronomical Bureau after the latter's death in 1688 —, also taught Kangxi. It seems that they were mainly in charge of teaching arithmetic and algebra, whereas the teaching of geometry was passed on to two of the five French Jesuits sent to China in 1685, namely Jean-François Gerbillon (1654-1707) and Joachim Bouvet (1656-1730).

Both teams of teachers produced textbooks in Manchu, which eventually were translated into Chinese. Bouvet and Gerbillon convinced the Kangxi Emperor that the using French textbook *Elemens de Géométrie* (1671) written by their confrère Ignace Gaston Pardies would make geometry easier to understand.[42] Since this textbook was in use in French Jesuit colleges when Bouvet and Gerbillon left for China, one may suppose that they felt more comfortable with it than with the Clavius edition of Euclid, which they may never had studied themselves.[43] During the same

[38] Standaert 2001: 719-721.
[39] Golvers 1993: 213 (n. 11).
[40] Han Qi 1999.
[41] Bernard-Maître 1945: nr. 458. Although this translation is no longer extant, Kangxi showed it to Bouvet and Gerbillon in 1690. Landry-Deron 1995, II: 38.
[42] Landry-Deron 1995, II: 38-39.
[43] Jami 1996.

period, Thomas produced, among others, a treatise of arithmetic, of which a Manchu version is still extant, and another of algebra, for which we only have the Chinese version. [44]

These treatises were in turn translated into Chinese under Kangxi's supervision. Since 1713, a number of scholars working at the newly set up Academy of Mathematics *(Suanxue guan)* revised them and included them in the *Yuzhi shuli jingyun* ("Imperially commissioned collected basic principles of mathematics"). The latter work was part of a wider compendium, the *Lüli yuanyuan* ("Source of Pitch-pipes and Calendar"), which also covered musical harmony and astronomy; it appeared under Kangxi's name in later bibliographies. This marked the final integration of Western learning into imperial learning, or, in other words, the absorption of Jesuit mathematical teachings into the body of knowledge which Kangxi bestowed upon his empire.

What should be taught

Throughout the period discussed above, the legitimacy and relevance of teaching the sciences in China were questioned both within and without the Society of Jesus. What was to be taught sometimes became controversial among those engaged in teaching. Such divergences were already apparent shortly after Ricci's death:

> *"In 1614 and 1615, F. Valentin Carvalho, provincial of Japan and China, commissioned* [Manuel Dias Jr. (1574-1659)] *to visit all the residences then existing, and to publish there the ban — soon to be lifted — on teaching to the Chinese mathematics or any other science, except that of the Gospel."* [45]

Carvalho never entered China. Although Dias is not known to have been versed in mathematics, neither he did limit his teaching to "the science of the Gospel". In fact, his *Tianwen lüe* ("Epitome of Questions on the Heavens", 1615-16) was printed shortly after Carvalho's order. A short account of the Ptolemaic cosmology, the work ends with a brief presentation of the newly invented telescope and the discoveries Galileo had made with it. This was a long way away from the "science of the Gospel".

---

[44] Standaert 2001: 740.

[45] Pfister 1932-34: 106.

Not only the teaching, but the very practice of the mathematical sciences in China were controversial, especially when in the service of the emperor. The activities of Johann Adam Schall von Bell (1592-1666), the initiator of Jesuit astronomy at the Astronomical Bureau, remained a contentious issue throughout his life. Some of his fellow Jesuits doubted that the vows taken by Jesuit priests were compatible with his acceptance of a charge in the Chinese civil service. In more general terms, they challenged his engaging in the sciences instead of devoting his energy solely to proselytising. [46]

Although hopes of converting the Kangxi Emperor were expressed and widely advertised by the French Jesuits at the beginning of their mission in China, the teaching of the sciences gradually changed status. To Ricci and his immediate successors, the sciences were a means to bring some Chinese scholars to conversion, because in their own education the sciences opened the way for human reason to approach the truths of religion. A century later, science had become a safe-conduct for the missionary enterprise, and for the French Jesuits, a means to construct a privileged relationship between their king and the emperor. In the 1710s, one of them, Jean-François Foucquet (1665-1741), defended the necessity of introducing recent astronomical innovations from Europe, against his confrères of the Astronomical Bureau who proposed maintaining the status quo in the calendrical methods used there:

> *"Father Foucquet[47] said on this matter that, by wanting to defend the known defects of Tychonic astronomy,[48] one could not but succumb; that on the contrary if the Europeans, managing to agree amongst themselves, unanimously substituted it with the theory founded on recent observations, if they declared for example that the paths of planets were elliptic or close to ellipses as Father Foucquet has shown it in his writings,[49] that would be enough to disconcert the servants of the Third Prince,[50] and to reduce them once more to the position of students of the Europeans; that with the six first books of Euclid and a few propositions from*

[46] Romano forthcoming.

[47] Foucquet refers to himself in the third person.

[48] The one in use at the Astronomical Bureau.

[49] On Foucquet's *Lifa wenda*, see Hashimoto and Jami 1997.

[50] Yinzhi, Kangxi's third son, was then in charge of the Academy of Mathematics mentioned above.

Archimedes on solids that have since been added, however well they might know them, it was not to be feared that they could penetrate by themselves the secrets of conic sections, or a more elevated geometry, that the modern theories of the Sun, the Moon and the other Planets would be a sealed language to them [...]" [51]

Like Smogulecki, Foucquet has been credited of attempting to introduce Copernicanism into China.[52] He certainly did mention Copernicus' heliocentric hypothesis in the astronomical treatise that he wrote for Kangxi. But his "theory based on recent observations" was a slightly modified Tychonic model: he took the five planets' orbits to be elliptic instead of circular.[53] In the above passage of a letter he sent to Rome, he was presenting this choice as a device for securing the position of missionaries in China. At the time the *Lüli yuanyuan* was being compiled at the Academy of Mathematics, headed by Kangxi's third son. This compilation was a means to reconstruct imperial science independently from the Jesuits, and to appropriate Western learning to the extent that the Western teachers would no longer be needed. Foucquet's response was to introduce a new set of results, many prerequisites of which were still unknown in China. Rather than teach the European sciences, his aim seems to have been to keep the emperor and his astronomers aware of their ignorance and dependence on the Jesuits. He was displaying his expertise without intending to share it.

Epilogue: Textbooks, but no teachers

In the above, we have followed some Jesuits at work teaching the sciences, at various times between the 1580s and the 1710s, but also to various audiences in various places: Beijing, Jiangnan and Fujian, Nanjing, the Manchu court. What was taught and how it was taught, the goals of the teachers and students, varied greatly not only according to the Jesuits' training, but also to the social status and political purposes of their students.

Since the dynastic transition, all the Chinese and Manchu scholars versed in mathematics and astronomy were aware of and willing to study Western learning. But, as we have seen, the Jesuits' progress towards the

---

[51] Foucquet 1716: 64.
[52] Witek 1982: 181-189. Martzloff 1994.
[53] This was the model adopted by the Jesuit astronomer Riccioli in his *Astronomia Reformata* (1665). Hashimoto and Jami 1997: 179-183.

centre of power lead them first to undertake the calendar reform in the 1630s, then to take office at the Astronomical Bureau, and finally to adapt their teaching to their imperial patron's needs and interests. As a result, from the dynastic transition on, most scholars who studied Western learning used the Jesuits' textbooks, without ever receiving their teaching. Qu Rukui and Xu Guangqi were among the very few students of Euclidean geometry who learnt it in conditions comparable to those which had studied the subject in Rome. That is to say, the textbook was an aid to the teacher's guidance. In contradistinction, most of those who wrote commentaries on the *Jihe yuanben* during the second half of the seventeenth century had studied it by themselves.[54] Their commentaries, rewritings, and syntheses with earlier extant Chinese works resulted in the integration of Jesuit textbooks into the corpus of the Chinese mathematical sciences. As to the manuscripts written for the emperor, they were circulated widely only after they had been edited by the Chinese and Manchu scholars in his service. This editorial work turned them into pieces of imperial scholarship, but also merged into a newly reconstructed mathematical corpus. In both cases there were two steps: the first one, that of translation, had been in part the work of Jesuits. The second one, that of appropriation, was entirely out of their hands.

REFERENCES

Baldini, Ugo. 2000. "The Portuguese Assistancy of the Society and Scientific Activities in its Asian Mission Until 1640," in Luis Saraiva ed., *História das ciências matematicas. Portugal e o Oriente*. Lisbon, Fundação Oriente: 49-104.

Bernard, Henri. 1935. *Matteo Ricci's Scientific Contribution to China*. Transl. Edward Chalmer Werner. Peiping: Henri Vecht.

Bernard-Maître, Bernard. 1945. "Les adaptations chinoises d'ouvrages européens: Bibliographie chronologique. I. Depuis la venue des Portugais à Canton jusqu'à la mission française de Pékin (1514-1688)." *Monumenta Serica* 10: 1-57; 309-388.

Billeter, Jean-François. 1979. *Li Zhi, philosophe maudit (1527-1602). Contribution à une sociologie du mandarinat chinois à la fin des Ming*. Genève: Droz.

[54] Engelfriet 1998: 362-431 discusses several of them.

Elman, Benjamin A. 1998. "The 'Chinese Sciences' in Policy Questions from Confucian Civil Examinations During the Late Ming," in R. Malek ed., *Western Learning and Christianity in China. The Contribution and Impact of Johann Adam Schall von Bell, S.J. (1592-1666)*. 2 vols. Nettetal: Steyler Verlag: 619-660

Engelfriet, Peter. 1998. *Euclid in China. The Genesis of the First Translation of Euclid's Elements in 1607 and its Reception up to 1723*. Leiden: Brill.

Foucquet, Jean-François. 1716. *Relation exacte de ce qui s'est passé à Péking par raport à l'astronomie européane depuis le mois de juin 1711 jusqu'au commencement d'octobre 1716*. Manuscript, ARSI Jap.Sin. II 154.

Gernet, Jacques. 1982. *Chine et christianisme. Action et réaction*. Paris: Gallimard.

Golvers, Noël. 1993. *The Astronomia Europaea of Ferdinand Verbiest, S.J.* (Dillingen, 1687). *Text, Translation, Notes and Commentaries*. Nettetal: Steyler Verlag.

Han Qi. 1999. "The Role of the Astronomical Bureau in the Catholic Mission during the Qing Period." In Noël Golvers ed., *The Christian Mission in China in the Verbiest Era: Some Aspects of the Missionary Approach*. Leuven: Leuven University Press: 85-95

Hashimoto, Keizo. 1988. *Hsü Kuang-Ch'i and Astronomical Reform. The Process of the Chinese Acceptance of Western Astronomy 1629-1635*. Osaka: Kansai University Press.

Hashimoto, Keizo and Jami, Catherine. 1997. "Kepler's Laws in China: A Missing Link? Jean-François Foucquet's *Lifa wenda*", *Historia Scientiarum* 6-3: 171-185.

Huang Yi-long. 1994. "Qu Rukui (Taisu) jiashi yu shengping kao," *Dalu zazhi* 89, 5: 8-10.

Jami, Catherine. 1992. "Rencontre entre arithmétiques chinoise et occidentale au XVIIe siècle", in P. Benoît et al. eds., *Histoire de Fractions, Fractions d'histoire*. Basel, Birkhaüser: 351-373.

Jami, Catherine. 1996. "From Clavius to Pardies: The geometry transmitted to China by Jesuits (1607-1723)", in F. Masini ed., *Western Humanistic Culture Presented to China by Jesuit Missionaries (XVII-XVIII centuries)*. Rome, Institutum Historicum S. I.: 175-199.

Jami, Catherine. 1997. "Giulio Aleni's contribution to geometry in China: the *Jihe yaofa*." In Tiziana Lippiello & Roman Malek eds., *"Scholar from the West". Giulio Aleni S.J. and the Dialogue between Christianity and China*. Nettetal: Steyler Verlag: 553-572.

Landry-Deron, Isabelle. 1995. *Les leçons de sciences occidentales de l'empereur de Chine Kangxi (1662-1722). Textes des journaux des Pères Bouvet et Gerbillon.* 2 vols. Mémoire de l'EHESS, Paris.

Li Jiubiao. *Kouduo richao.* BNF Chinois 7114.

Martzloff, Jean-Claude. 1994. "A Glimpse at the post-Verbiest Period: Jean-François Foucquet's *Lifa wenda* (Dialogue on Astronomy) and the Modernization of Chinese Astronomy or Urania's Feet Unbound." In John Witek ed., *Ferdinand Verbiest, S.J., Jesuit Missionary, Scientist, Engineer and Diplomat, 1623-1688.* Nettetal, Steyler Verlag: 519-529.

Pfister, Louis 1932-34. *Notices biographiques et bibliographiques sur les Jésuites de l'ancienne mission de Chine, 1552-1773.* 2 vols. Shanghai: Imprimerie des Lazaristes.

Ricci, Matthieu, & Trigault, Nicolas. 1978. *Histoire de l'expédition chrétienne au royaume de la Chine (1582-1610).* Paris: DDB/Bellarmin.

Romano, Antonella. Forthcoming. "Etre jésuite en Chine au milieu du XVIIe siècle: polémique sur le Tribunal des Mathématiques de Pékin."

Shi Yunli. 2000. Tianbu zhenyuan yu Gebaini tianwenxue zai Zhongguo de zhaoqi chuanbo. *Zhongguo keji shiliao* 21,1: 83-91.

Siku quanshu zongmu. 1981. 2 vols. Beijing: Zhonghua Shuju.

Standaert, Nicolas, ed. 2001. *Handbook of Christianity in China.* Vol. 1: 635-1800. Leiden: Brill.

Standaert, Nicolas. 2001. "Xue Fengzuo's and Smogulecki's Translation of Cardano's Commentaries on Ptolemy's *Tetrabiblos*", *Sino-Western Cultural Relations Journal* 23, pp. 51-80.

Wang Zhongmin ed. 1984 *Xu Guangqi ji.* Shanghai: Shanghai guji chubanshe.

Witek, John. 1982. *Controversial Ideas in China and in Europe: A Biography of Jean-François Foucquet, S.J. (1665-1741).* Rome: Institutum Historicum S.I.

Zürcher, Erik. 1997. "Aleni in Fujian , 1630-1640: The Medium and the Message." In Tiziana Lippiello & Roman Malek eds., *"Scholar from the West". Giulio Aleni S.J. and the Dialogue between Christianity and China.* Nettetal, Steyler Verlag: 595-616.

NEWS FROM CHINA IN SIXTEENTH CENTURY EUROPE: THE PORTUGUESE CONNECTION

Rui Manuel LOUREIRO
Universidade Lusófona, Portimão

Before the discovery of the Cape route to India in 1498 by Vasco da Gama, little was known in Portugal, or indeed in Europe, about East Asia. European literary circles had some knowledge of those distant places, of course, based on information collected and spread by mediaeval travellers like Marco Polo who had crossed Asia or by compilers like John de Mandeville who had gathered all the material available on the non-European world. But this literary information, which reached only a small number of people, presented a somewhat vague and sometimes fantastic image of the real East Asian world. The cartographic representation of Asian lands available to the learned European public in mediaeval maps was also rather imaginative and symbolic, bearing little or no relation to the real world, as a quick glance at the *Catalan Atlas*, prepared around 1375 by Abraham Cresques, will show.

After Vasco da Gama arrived in Western India by the sea-route, in the last years of the fifteenth century — almost exactly five hundred years ago — everything suddenly began to change. Portuguese ships rapidly came into contact with most of Asia's maritime shores, from the Red Sea, visited as early as 1503, to the remote islands of Japan, where they anchored for the first time about forty years later. In the course of the sixteenth century, Portuguese navigators were responsible for the establishment of direct and regular relations between East and West, and also for the spread all over Europe of accurate and detailed geographic and ethnographic information about the distant and previously unknown, or little known, lands of Asia.

A Portuguese author who wrote a *Tratado das coisas da China* ("Treatise on Things Chinese"), published in Évora in 1570, reflected upon the enormous intellectual revolution brought about by the Portuguese voyages of discovery and exploration. Commenting on ancient stories and legends about pygmies who fought with griffins for the possession of gold, and about men who had very small mouths and sucked their food through straws, and about other men who had an enormous foot which gave them shade when lifted up over their heads, he wrote:

> *"These and other things which they affirmed of those parts* [of Asia], *were proved to be fables after that India had been discovered by the Portugals."*

Throughout the sixteenth century, the Portuguese kept a tight control of the Cape route to India, successfully opposing the arrival of other European ships on the Oriental seas. At the same time, they established the grounds of a loosely organised but highly efficient political-administrative-military body known as the ***Estado da Índia***, which was composed by a string of forts and factories built along the Asian shore-line, linked together by powerful fleets. The ***Estado da Índia*** — whose strategic points were Hormuz, on the Persian Gulf (occupied in 1507), Goa, on the west coast of India (conquered in 1510), Malacca, on the Malay Peninsula (conquered in 1511), and Macao, on the South China coast (occupied after 1557) — had as its main purposes a systematic involvement in the most profitable Asian trading networks and the diversion to Lisbon of a significant share of the trade in luxury commodities. For almost a century, the Portuguese were the only European power established in the East, and they controlled a large part of the spice and drug trade to Europe, and also some of the most important regional trade routes in Eastern waters. As they controlled the direct sea route between Asia and Europe, the Portuguese also controlled the Western information supply on Eastern matters. A significant part of the news about Asia, and about China, of course, available in sixteenth century Europe came by way of the Portuguese. We can therefore speak about a ***Portuguese connection*** in the construction of European perceptions of China's realities.

China, for that matter, was an altogether unknown geographic concept in Europe, because the regions we now know by that name were called ***Cathay*** and ***Mangy*** in mediaeval literature and cartography, thanks to the reports of Marco Polo and other travellers. But in the wake of Vasco da Gama's arrival in India, information gathered by the Portuguese was to lend increasing substance to the concept of China, contributing to the dissemination throughout Europe of an increasingly accurate and detailed image of the ***land of the Chins***. Step by step, China revealed itself to the eyes of Portuguese travellers, as a fabulously rich and powerful kingdom, which played a central role in the political, economic and cultural world of East Asia. The word ***China***, by the way, seems to have been derived from ***Chin***, the name of an ancient Chinese dynasty which became known in India, Persia and other regions of Western and South Asia immediately before the Christian era. The final **a** is thought to have been added by the Portuguese, who picked up the place-name ***Chin*** in the ports of Hindustan.

Various types of sixteenth century sources circulated under Portuguese supervision and were responsible for bringing the Chinese world to the attention of Europe. Information was diffused, first of all, by ***oral reports***, impossible to quantify; but we know that, besides Portuguese, other European merchants, sailors and craftsmen travelled to and from China on board Portuguese ships, on a regular basis. Many of them, once they were back in Europe, contributed to the dissemination of knowledge about ***things Chinese***. Information about China was also spread through the so-called ***silent sources***: porcelain objects; silks; wooden furniture; plants like china-root, rhubarb and tea; printed books in Chinese ideograms; animal products like musk; paintings on silk or rice-paper; and so on. These exotic products, brought to Lisbon in huge quantities, were immediately shipped to European ports in the Low Countries and in the Mediterranean, where they were extremely important in the process of diffusion of information about Chinese matters, since they conveyed to the European public new images of a distant and previously unknown world. ***Literary sources*** were, of course, the single most important factor in the construction and diffusion of the European image of China. Written reports and sketch maps produced in the East by Portuguese observers with considerable overseas experience, as well as by home-based scholars in Lisbon, circulated widely, both in manuscript and printed form, and were responsible for the spread, throughout Europe, of new data about all ***things Chinese*** and also about the particular circumstances of Sino-Portuguese relations.

The earliest information compiled by the Portuguese, naturally enough, had to do with matters pertaining to commerce and navigation, for, in order to intervene in the trading world of Asia, it was imperative to know where products came from and how to get there, through what means they came and at what time of year, how much they cost and what could be offered in return. Matters of economic geography, then, had the highest priority.

But political and religious information was also relevant, because it was necessary to find out which states would welcome or tolerate the Portuguese as a new trading partner, and where it was possible, and by what means, to establish permanent bases. In this context, the oldest explicit reference to China appears as early as 1502, in a note on the so-called ***Cantino*** World-map, placed near Malacca:

> *"In this city there is every kind of merchandise that comes from Calicut, to wit, cloves and benzoin and lignaloes and sandalwood, styrax and rhubarb and ivory and most valuable precious stones and musk and fine porcelain and much other merchandise; all of these, for the most part, come from abroad, from the Land of the Chinese."*

China immediately became the focus of great interest in the plans of Portuguese statesmen, and quite soon steps were taken to reach its southern shores. After the conquest of Malacca, in Southeast Asia, by a Portuguese fleet under the command of Afonso de Albuquerque, in the year 1511, the gateway to the Far East lay wide open, because the former Muslim sultanate maintained regular contacts, which could now be renewed, with the more important trading centres of East Asia, including South China. Two years later, in 1513, Portuguese navigators visited the islands in the Pearl River delta, establishing for the first time a direct maritime link between Europe and the Middle Kingdom, only fifteen years after Vasco da Gama's arrival in Calicut. Detailed maritime charts, along with optimistic trade reports, were immediately sent back to Lisbon, increasing the Portuguese interest in ***things Chinese***. A brief navigational guide then prepared by Francisco Rodrigues included basic instructions to ships sailing from Malacca to Canton and back. Sino-Portuguese relations had been opened, never to be closed again.

Between 1515 and 1516, in response to Europe's enormous curiosity about the Eastern world, Tomé Pires and Duarte Barbosa, two Portuguese civil servants with considerable overseas experience, prepared two global geographic treatises on Asian matters, the *Suma Oriental* and the *Livro das coisas do Oriente*, respectively. Persistent enquiries undertaken by the Portuguese in the ports of Asia had paved the way for the accumulation of much information about lands and seas, routes and products, peoples and societies, economies and cultures, information that was now, for the first time, compiled in a systematic manner. Though they were not printed in Portugal at the time, manuscript copies of these treatises achieved reasonable circulation throughout Europe, and were eventually published in Venice, in 1550, by Giovanni Battista Ramusio, in the first volume of his celebrated *Navigationi et Viaggi*.

Obviously, both works devoted considerable space to China, which was considered an especially attractive region, on account of its immense riches. The pharmacist Tomé Pires, alongside important data on available merchandise and trade conditions, recorded innovative information on the tributary system that regulated China's foreign affairs and on Beijing's official closed-doors policy. According to this author, the Portuguese ought to pay special attention to "the city of Canton", because this was "the place where the entire kingdom of China unloads all its merchandise, both from the hinterland and from the sea". Duarte Barbosa, in his turn, underlines the vastness of the "great kingdom of China" and the enormous power of its supreme ruler, elaborating also on valuable merchandise of Chinese origin, with a special emphasis on porcelain, the nature of which he describes in detail, though with a generous portion of imagination.

After reaching the South China coast, the Portuguese soon tried to establish a permanent base there, as they had done in other parts of Asia, from where they could organise regular trading relations with the west coast of India, where the ***Estado da Índia*** had its central base. But, unlike in other Asian maritime regions, where the Portuguese easily found conditions for their permanent settlement, through peaceful compromise or military intimidation, this enterprise proved to be somewhat more difficult on the South China coast. Four decades were to pass before the Portuguese could find a safe haven on Chinese territory. The deliberate policy of isolation implemented by the Chinese court in the first half of the sixteenth century certainly contributed to the failure of formal Portuguese approaches, but the overbearing attitude and enormous lack of tact shown by the Portuguese navigators were also responsible for an initial period of official confrontation between Portugal and China.

Regardless of state-to-state relations, Portuguese and Chinese merchants soon established informal partnerships, based on a mutually profitable trade, carried out on deserted islands off the Chinese coast. Ships coming from Malacca regularly visited the Chinese coasts of Guangdong, Fujian and Zhejiang provinces, exchanging Southeast Asian pepper and incense for Chinese silk and porcelain. The Portuguese ***discovery*** of Japan, around 1542, contributed greatly to the increase of business transactions on the South China coast, because relations between the Middle Kingdom and the Land of the Rising Sun were by then suspended, on account of previous violent incidents caused by the Japanese on eastern Chinese ports. The Portuguese navigators, on learning of this situation, rapidly became indispensable middlemen in the trade in Chinese silk and Japanese silver.

The increase of Portuguese trading expeditions on the South China coast, with its inevitable benefits for the development of the southern provinces' economies, soon led to an agreement between the Chinese authorities and foreign merchants in the region of Canton. By 1554 the Portuguese were allowed to trade freely on the islands of the Pearl River delta, as long as they paid proper customs duties, and could even go up to the Canton fairs. Three years later they had a permanent settlement in the Macao peninsula. It is logical that the Cantonese mandarins would seek to concentrate foreign trade on a single point of Guangdong's long coastline, in order to reduce contraband and piracy, while, at the same time, securing a new source of income. The development of Macao was so fast that by 1564 there were already about 800 Portuguese permanently living there, alongside twice as many Asians. With the blessing of the Chinese, the Portuguese soon transformed Macao into a major international port, thanks to their extensive business connections and their intimate involvement in the South China Sea trading networks. Regular links were established with Malacca and Japan, but also with Siam, Cambodia, Vietnam, the Philippine Islands and the Malay Peninsula.

Closer trading relations with China also meant better opportunities to observe Chinese realities and to collect information about day-to-day life in China. So, Portuguese voyages along China's maritime borders during the 1540's and 1550's coincided with a steady flow of written reports on ***things Chinese*** and of detailed maps of the Chinese coast. The Portuguese began by paying attention to facts of a more utilitarian nature, involving the physical environment and the essential world of trade. After filling the more urgent information gaps, Portuguese travellers expanded their questionnaires, seeking to secure more detailed information concerning habits and customs, systems of government and forms of administration, religious beliefs and practices, town planning and manufacturing systems, in order to obtain a more accurate overall picture of the Flowery Kingdom. The importance of the Jesuits in this context must be emphasised, because these missionaries, present on China's coast from 1552 on, played a major role in the formulation of cultural inquiries, which were filled by Portuguese and Chinese informants.

Among other sources, Galiote Pereira's *Tratado da China*, written around 1552, in the wake of a period of imprisonment in the southern provinces of China, warrants special attention. Most impressed with what he had seen, this writer stressed the huge size of the Middle Kingdom and the infinity of its population. He also openly praised many aspects of Chinese life, such as its perfect highways and bridges, the impeccable layout of the cities, the rational organisation of economic production, the efficiency of local administration and the impartiality of the judicial machinery. In his own words, in China "everything goes so well that it can truly be said that it is the best-governed land that there can be in the whole world". And he went on, with tremendous openness, to say that the Chinese "had little reason to envy" the Portuguese, for in some aspects it could be said that Portugal would lose out in an open confrontation with this remote Far Eastern empire. Though it was not fully printed in Portugal at the time, manuscripts of Galiote Pereira's short treatise were widely read. The information it contained was to make a decisive contribution to Portuguese knowledge of Chinese civilisation, both confirming and expanding the information that had been provided by some earlier writers. Pereira's work was quickly translated into other European languages, and soon appeared in print in Italian (Venice, 1565) and in English (London, 1577).

Based on reports written by Galiote Pereira and others, which began to arrive in Portugal about 1550, the great Portuguese chroniclers of the sixteenth century gave special attention to ***things Chinese***. Fernão Lopes de Castanheda, for instance, included a detailed description of China in one of the volumes of his *História do Descobrimento e Conquista da Índia pelos Portugueses*, which appeared after the middle of the century (Coimbra, 1552-1554). Lopes de Castanheda, who was a man of considerable oriental experience, described, among other matters, Chinese Buddhism, the structure of Chinese government and the mechanisms of local power, providing a wealth of detail on these subjects. His vision of China was extremely positive: "The Chinese are men of singular ingenuity, both in the liberal arts and in the mechanical". And he then went on to say that: "Among them is used all the polity in the world, and they consider that there is none in any other part than China, nor do they hold as men those who are not Chinese". Lopes de Castanheda's monumental work was widely translated into other European languages, namely into French (Paris, 1553; Antwerp, 1554), Spanish (Antwerp, 1554), Italian (Rome, 1556; Venice, 1577-1578) and English (London, 1582).

João de Barros, the official chronicler of Portuguese overseas expansion, whose first three *Décadas da Ásia* were printed in Lisbon between 1552 and 1563, also paid particular attention to Chinese matters, expressing great admiration for the Chinese civilisation. Thus, he said in one of his chronicles:

> *"With regard to the king of China, we may well declare that only he, in land, people, power, riches and polity, is greater than all."*

João de Barros never actually visited Asia, but so strong was his interest in China that he went as far as to buy an educated Chinese slave to translate such Chinese books as he was able to get his hands on. Praising all Chinese undertakings in the field of material life, João de Barros said in conclusion that "in these people we find all things that are praiseworthy in the Greeks and Romans alike", a somewhat surprising statement coming from a humanist of his standing. João de Barros' *Décadas* were soon translated into other languages, circulating widely in European literary circles. The second edition of Giovanni Battista Ramusio's ***Navigationi et Viaggi*** (Venice, 1554) included Italian extracts of the Portuguese chronicler's work on Asia. In 1562 the first two *Décadas* came out in Italian translation in Venice.

Fernão Lopes de Castanheda's *História* and João de Barros's *Ásia*, together with Galiote Pereira's *Tratado*, gave rise, in Portugal, to a fascinating process of knowledge and critical idealisation of Chinese reality. China, for the Portuguese, became a marvelous place, concentrating all the characteristics of a model society, in political, economic, technological, administrative, judicial and even intellectual terms. This interest in and admiration for the Flowery Kingdom, governed by "one of the greatest known kings of the world", as one Portuguese author wrote in 1563, found echoes in all the great Portuguese literary works dedicated to overseas matters during the second half of the sixteenth century. To name some famous examples, we can find long references in praise of China in the *Comentários de Afonso de Albuquerque*, prepared by Brás de Albuquerque (Lisbon, 1557 and 1576); in António Galvão's *Tratado dos Descobrimentos* (Lisbon, 1563); in the *Colóquios dos Simples e Drogas da Índia* by Garcia de Orta (Goa, 1563); in the *Crónica do Felicíssimo Rei D.Manuel* by Damião de Góis (Lisbon, 1566-1567); in Jerónimo Osório's *De rebus Emmanuelis gesti*, (Lisbon, 1571); and in many other works.

Through these authors, China became a major cultural interest for cultivated Portuguese, as a far distant Asian kingdom where the undertakings of mankind in the fields of administration, production, city planning, justice, law and order, and civil manners warranted the greatest attention as potential models for the European world. Many of these works found their way, directly or indirectly, into the most important European cultural centres, spreading information about China. Garcia da Orta's botanical treatise, for example, was widely circulated through several reprints of a Latin translation made by the Flemish humanist Charles de l'Écluse (Antwerp, 1567). And the Latin chronicle of king D.Manuel I, by Jerónimo Osório, was reprinted countless times all over Europe in the years following its first publication.

A single shadow darkened the immensely positive image that Portuguese writers presented of China: the religious question. In Portuguese sixteenth century literature, every aspect of Chinese civilisation is praised without limits, with the exception of religious practices and beliefs. A passage from Galiote Pereira's *Tratado* expresses in an admirable manner the fundamental importance then given to religion: according to this writer, Portuguese judges, "save the wand of office of each one, could very well serve" the Chinese mandarins, "were it not for their being heathen, for clearly a Christian cannot be put to serve a heathen". For sixteenth century Portuguese, the religious status, more than any other aspect, marked the fundamental difference between the civilisations.

The process of knowledge and idealisation of the Chinese world that developed in sixteenth century Portuguese literature culminated in the *Tratado das coisas da China*, prepared by Gaspar da Cruz (Évora, 1570), which was the first monograph devoted solely to China to be published in Europe. The Portuguese priest sojourned briefly in the islands of the Pearl River and in Canton, just a year before the founding of Macao. On his return to Portugal, Gaspar da Cruz drew up a very successful summary of all the available knowledge about the Middle Kingdom, based on his own recollections and also on oral and written information provided by Portuguese merchants and adventurers. He took care to collect data concerning every aspect of Chinese life capable of impressing Portuguese readers.

In the first place, his description exhaustively addresses matters such as geographic location, administrative boundaries and limits, productive activities, city planning, the civil service and the judicial and prison systems. The missionary then dealt with the physical aspect of the Chinese, their habits and customs, their festivals, music, writing, beliefs and many other aspects of everyday life. The work of the Portuguese missionary is so thorough that it even gives details about many characteristics of Chinese society that had gone unnoticed by earlier observers, such as the use of tea, Chinese writing, the importance of printing, the size of the Great Wall, the habit of binding women's feet, the practice of cormorant fishing, and indeed all the great topics that, down to modern times, were to mark the European vision of the Celestial Empire. Gaspar da Cruz presents an extremely favourable view of China, even accepting the superiority of certain aspects of the Chinese society in comparison with Portugal. These include the impeccable organisation of the cities, the quantity of roads and bridges, the rational use of the land, the ingeniousness of the artisans, the unquestioned authority of the mandarins, the utter honesty of certain senior civil servants, the low taxation and state charity. The balance of the *Tratado das coisas da China* is frankly a positive one, for, in the words of the Dominican father, "the Chinese exceed all the others in populousness, in greatness of the realm, in the excellence of polity and government, and in abundance of possessions and wealth".

The *Tratado das coisas da China* was immediately circulated outside Portugal, and in the course of time was to have an enormous impact on European knowledge of China. First, it was largely paraphrased by Bernardino de Escalante, a Spanish author who published in 1577, in Seville, a geographical treatise about China, with the curious title of *Discurso de la Navegacion de los Portugueses*. Then, a few years later, the works of Gaspar da Cruz and Bernardino Escalante were used by the Spanish missionary Juan González de Mendoza as sources for his best-selling *Historia de las cosas más notables, ritos y costumbres del Gran Reino de la China*, first printed in Rome in 1585, but subsequently published, in several European languages, in no fewer than forty different editions before 1600. González de Mendoza's *Historia* alone, was largely responsible for the diffusion of information about China throughout Europe, much of which had been systematically collected by the Portuguese since the 1550's.

The increased number of trade expeditions to the South China coast also led to the regular production of maps and navigational guides. During their incessant voyages, the Portuguese had the opportunity to make fairly accurate and exhaustive surveys of the topography and hydrography of maritime China. As a result, many manuscript navigational guides began to appear, containing precise indications about sailing conditions: the winds and currents, shallows and anchorages, and navigation calendars. Gaspar Moreira's *Livro de Marinharia*, compiled early in the seventeenth century, preserves some of these anonymous technical texts, such as the "Roteiro de Malaca para a China" ("Navigational Guide from Malacca to China") and the "Roteiro de Malaca para o Japão" ("Navigational Guide from Malacca to Japan"). Many other similar documents, written by unknown Portuguese pilots, were collected at the end of the sixteenth century by the Dutch traveller Jan Huyghen van Linschoten in his monumental *Reys-gheschrifft van de Navegatien der Portugaloysers in Orienten*, first printed in Amsterdam in 1595. With the help of these materials, Dutch and English ships were soon to find their own way to the South China Sea.

The Portuguese cartography of the times also registers the information revolution then under way. The outline of the Chinese coast was becoming more accurate and the place-names increasing in number, clearly bearing witness to greater familiarity with these remote areas as a result of ever more frequent voyages. Some of Diogo Homem's charts, for example, show this evolution. His manuscript *Atlas* of 1558 includes a chart of the Far East that shows a "China Provintia", the coast of which is set out with considerable precision. Several placenames are shown, such as ***Veniaga*** Island, ***Chinchéu***, ***Rui Lobo*** Bay, ***Lamau*** Island and ***Formosa*** Island. Another map by the same cartographer, included in a 1561 *Atlas*, repeats the same placenames and adds others such as ***Cabo dos Pescadores***, ***Aguada*** Port and ***Sal*** Island, which, in their uniqueness, bear witness to the considerable amount of Portuguese shipping, even though many of them still await positive identification.

A ***Mapa da Asia*** by Bartolomeu Velho, also drawn in 1561, confirms that the Portuguese had a great deal of knowledge of the topography of the southern provinces of China. Indeed, the royal cartographer entered no less than forty place-names along the coast of the Celestial Empire, from ***Sanchoão***, in the West, to the ***Rio de Liampó***, in the East. But the cartographic representation of China in the sixteenth century reached its zenith with Luís Jorge de Barbuda. This Portuguese cartographer, around 1575, drew a detailed map of the entire Chinese territory, based on information provided by his fellow countrymen and perhaps with the aid of some Chinese maps to which he may have had access. This map was published about ten years later, under the title "Chinae, olim Sinarum regiones, nova descriptio", in the Amsterdam-printed 1584 edition of the monumental work *Theatrum Orbis Terrarum*, compiled by Abraham Ortelius. For the first time, a Portuguese chart included the names of all of China's provinces and the names of many inland cities, as well as an approximate representation of the Chinese hydrographic system. The coast of the Celestial Empire was scattered with place-names obviously of Portuguese origin.

Throughout the sixteenth century, the Portuguese led the way in the complex process of European approaches to China. On the one hand, they were the first representatives of Europe to establish direct links by sea with this far distant empire, from as early as 1513. Their nearest competitors in the Far East, the Spanish, were only to provide a regular link between Mexico and the Philippines after 1565, when Andrés de Urdañeta finally discovered the northerly return route to the New World. Relations between Portugal and China, which after 1557 were undertaken via Macao, led to increased interaction at every level, allowing Europe to be flooded with hitherto undreamt-of quantities of silks, porcelain and other products from the Celestial Empire. On the other hand, the Portuguese played a fundamental role in the ***discovery*** of the Chinese civilisation. Not restricting themselves to simple trading, they were also concerned with the gathering and recording of detailed information about every aspect of the Chinese world. The information thus accumulated, was subsequently circulated, thanks to the wide international diffusion of treatises, chronicles, reports, maps and navigational guides of Portuguese origin. The apologetic nature of a large part of this mass of information was responsible for the development of a totally positive image of China in Portuguese culture in the second half of the sixteenth century.

The positive nature of the Portuguese image of China arose for four essential reasons:

a) First, the traveller's point of departure always affects the observations made: aware of the limitations of their own society, they were always on the look-out for aspects of China that could compare favourably with the European world. And, in this respect, the solutions found by the Chinese for certain problems of everyday life could not but cause admiration.

b) Secondly, the ***otherness*** of Chinese civilisation — with its huge spaces, its gigantic urban agglomerations and its truly infinite population — was such as to produce an irrepressible feeling of surprise in a European observer, particularly a Portuguese observer from a country that was truly minuscule in Chinese terms.

c) Thirdly, the great distance that separated Portugal and East Asia must surely have contributed to the enthusiastic manner in which the news concerning this Asian power were received. Physical distance removed any potential threat, allowing a more tranquil appraisal of the information that was received.

d) Lastly, the specific limitations that affected the experiences of Portuguese navigators introduced an element of distortion into their observations. Indeed, till 1583 the Portuguese had contact almost

> exclusively with the maritime regions of China, which were quite prosperous, and they did so always through the medium of interpreters, since they did not know the Chinese languages. And although they could, to a certain extent, trust their own perceptions, they were obliged to generalise on the basis of relatively limited experience. The Portuguese images of China, during a large part of the sixteenth century, were images of a certain part of the Chinese world, that could in no way be extended to the whole of the Middle Kingdom.

After 1583, Matteo Ricci and some of his fellow friars obtained authorisation from the Guangdong authorities to enter the Chinese mainland, thus beginning a new process of collecting information about every aspect of Chinese life — ranging from imperial administration to intellectual life, including social organisation and economic resources. The Jesuit missionaries were soon to discover another side to China, immediately after acquiring an intimate knowledge of the Chinese language, during their longer or shorter travels through the provinces of the interior. From their strategic base in Macao, they were to make a decisive contribution to the widening and deepening of European perceptions of the Celestial Empire. In effect, the priests of the Society of Jesus took on the role of official informers, as they channelled to Portugal an inexhaustible quantity of factual data, of which regular use was made by Portuguese overseas literature. And these written materials were immediately conveyed to Europe, either through Jesuit channels or through scholarly circles.

The very long treatise *De Missione Legatorum Japonensium*, coordinated by Duarte de Sande, and printed in Macao in 1590, bears witness to the enormous changes that were taking place at the century's end in respect to European approaches to China. This work, which set out to narrate the journey made to several European countries by four young Japanese ambassadors, included at one point a lengthy chapter on China, in which details were given based on the information recently acquired by Matteo Ricci in the Chinese mainland. Important shifts began to take place in the European image of China. But that is another story altogether, one that cannot be explored here. It is worth mentioning, to conclude, that a copy of this Latin Jesuit travelogue was on board the Portuguese carrack ***Madre de Deus***, which was captured in the Atlantic, off the islands of the Azores, in 1592, by English privateers. Richard Hakluyt, the famous British geographer and compiler of travel literature, immediately prepared an English translation of the chapter on China, including it in one of the volumes of the second edition of his *Principal Navigations*, published in London in 1598-1600. This last example, once again, shows the extraordinary importance of the ***Portuguese connection*** in European access to knowledge about China.

ESSENTIAL BIBLIOGRAPHY

Boxer, Charles Ralph, *South China in the Sixteenth Century*, London: The Hakluyt Society, 1953.

Lach, Donald F., *Asia in the Making of Europe – The Century of Discovery*, 2 bks., Chicago: The University of Chicago Press, 1994.

Loureiro, Rui Manuel, *Um Tratado sobre o Reino da China dos Padres Duarte Sande e Alessandro Valignano (Macau, 1590)*, Macao: Instituto Cultural de Macau, 1992.

Loureiro, Rui Manuel, *Fidalgos, missionários e mandarins – Portugal e a China no século XVI*, Lisbon: Fundação Oriente, 2000.

English revision by Paul Covill.

THE INDIANIZATION OF SPAIN IN THE XVITH CENTURY

Juan GIL
University of Seville

The aim of this study is to establish the extent of the Indian influence — via Portugal — on Spanish civilization during the XVIth century. This is the first of a series of three papers: the subject of the other two will be the influence respectively of China and Japan on XVIth century Spain. The epochal book of Lach is based mainly on printed sources, so that the reader may get the wrong impression of the real influence of Asia on XVIth century Europe.

True knowledge of a country is only acquired by the study of its languague, literature and culture. However, very few Spaniards (e.g. Saint Francis Xavier) had any idea of the Oriental languages; but even these few Spaniards only knew the spoken language; they could neither read nor write it. Moreover, the European mentality in the XVI century forbade any real approximation to Oriental culture: the Europeans were fully convinced of their own superiority, for they were Christian and they brought with them, besides civilization, as they thought, the true religion and the true god. It was thanks precisely to the religious mission that the books that opened Asia to Europe were published: the treaties of Gaspar da Cruz, Juan González de Mendoza, Ricci-Trigault, the Jesuit *annuae*, etc. The merchants (Tomé Pires, Duarte Barbosa) indeed wrote invaluable reports about the commerce and the maritime and terrestrial roads that linked the markets of the East with the markets of the West; but these reports were not intended to be published, and if they were printed it was due only to the diligence of G.B.Ramusio. The knights and the soldiers wrote to praise their own deeds or to exalt national achievements. Even the Renaissance geographers tried to adapt the new discoveries to the Ptolemaic doctrine and the old myths: Cattigara was identified with the island of Gilolo by Tomás Durán,

Sebastian Cabot and J. Vespucci; Waldseemüller depicted in the North-Eastern corner of Asia the home of the *Judei clausi*, and even Ortelius searched for the Ten Lost Tribes in China and Mongolia.

For the European, the Far East was a land of pagans to be converted or conquered, a land in consequence to be visited by soldiers or by friars. If the existence of an intellectual influence on the Spanish culture of the XVIth century must be discarded, however, India exercized a powerful attraction on the people of Spain through the wares exported by the Portuguese armadas, wares that are extremely illustrative of the Indian technology. It is in this sense only that we can speak of Indianization of Spain. In order to estimate the true importance of this Indianization, we must deliberately exclude from our study the following wares:

a) The properties of the men who lived in India or had some link with it. Moreover, a man can be well informed about the Orient, but not his fellow-countrymen. What we are trying to measure is the Indian impact on the mass-media, not on one single person.

b) The Oriental curiosities belonging to kings and their relatives. We know, for instance, that the sarcophage in which Philip the II was buried in the Escorial was made of Indian wood, *angelí*; and we know also that the same king promised to his son as a present an Indian desk, probably adorned with ivory. But these cases, as can be easily understood, are irrelevant to our purpose.

c) We are not interested in the wares documented only once, or by the first and rare documentation of an Indian word. Let me give an example. The book of Duarte Barbosa was very early translated into Spanish: it was still the time when Portugal and Spain were rivals for control of the Spicery. In the translation of this book some Indian words appear, but I doubt if their meaning was completely understood even by the courtiers of Charles the First.

d) Always bearing in mind the ambiguity of the very term India, we must discard all the examples that could refer to the Occidental Indies. An example: the coconut was seen in Europe as a curiosity in the early XVIth century; from Portugal, Clenardus sent coconuts to Vasaeus, thinking that it was a pleasant and interesting gift for a professor at Salamanca. Coconuts were artistically put in silver frames, as can be seen in Museums. However, we must remember that coconuts were also brought from America.

e) Other objects may have an African origin: for instance, ivory or ebony. A very special case is the so-called unicorn, a very vague word that can also designate the tusk of the elephant. The viceroys of India sent to the kings of Portugal rhinoceroes. The name given to the smaller *rhinoceros Sondaicus*, 'bada' or 'abada', appear in Spanish texts at the end of the XVIth century, because Philip the II had an 'abada' in Madrid, where it was shown to the Japanese embassy in 1584. Pedro Pais saw the same abada in 1587. But this rhinoceros only served to satisfy the curiosity of the dwellers and visitors at Madrid and not to stimulate an Orientalist vocation.

Having enumerated the negative criteria, it is time to consider the other side of the question. The positive criteria can be reduced essentially to two. In fact, only two will be taken into account:

a) the wares that found prestige in Spain;

b) the wares whose Indian origin was recognised by the owners.

Let us take another example. Probably Queen Isabella possessed in her treasures some Indian curiosities: Dom Manuel was twice her son-in-law (he married first Isabel, then Maria) and it is very likely that he sent some Indian gifts to the parents of his wife after Vasco da Gama's triumphant voyage. However, when Queen Isabella died in 1504, nothing appeared in the auction of her properties that could definitively be called Indian, although we can attribute an Indian origin to a wooden box containing perfumes. It is, then, to be deduced that Queen Isabella was not conscious of having any Indian object, as, for instance, she was fully aware of possessing damask from Georgia.

The present study is based largely on the huge mass of documentation furnished by dowries, wills, inventories, etc. conserved in the Archive of Notarial Protocols at Seville. It is documentation not easy to read nor easy to collect, but it throws considerable light on the goods exported from India.

Before passing to the study of the documents, something must be said about the informants, that is the men that went to the Far East or the men that came from the Far East, the possible mediators between India and Spain. In the first years of the Indian run some Spaniards went to India with the Portuguese, some of them exiliated nobles who finally became Portuguese (Saldaña) or soldiers (the famous Guadalajara). In 1505, Pedro de Anaya was appointed by Dom Manuel captain of Sofala; with thim went many Spaniards, and one of them, Martín Fernández de Figueroa, wrote the first book published on the deeds of the Portuguese (and Spaniards) in India (Salamanca, 1515). The Burgalese merchants sent ships in the Portuguese fleet (the "nao Burgaleza d'armadores"), earning a lot of money (100% according to the documentary sources). To the soldiers and the merchants we must add the friars: one of them, the Dominican father Caro, won some fame in India.

We must also take into account the Portuguese who lived in Spain. A typical case is that of Diogo Barbosa, the servant of D. Alvaro de Portugal (and afterwards of his son D. Jorge). When Dom Manuel allowed D. Alvaro to send a ship to India, the captain of the caravel was Diogo Barbosa, who, after his Indian voyage, returned to Seville where he remained until his death. Very different was the life of Simâo de Tovar (+ 1597), the friend of Arias Montano, C. Clusius and B. Paludanus: he did not go to India, but he had a great deal of information about the Far East and possessed Indian wares, among them a rhinoceros horn (*abada*).

The wares

The Portuguese ships returned to Lisbon loaded mainly with pepper. We know that some pepper came from Lisbon to Seville because the cargo of a Portuguese ship was seized by the Inquisition in 1511, on the claim that the money was loaned by "heretics", viz. Spanish-Jewish merchants. As pepper was used before 1499, there was no great innovation in the trade. However, the Portuguese brought to the Iberian Peninsula other wares that found great acceptance. Cloth especially was a very appropriated merchandise to fill out the cargo.

A) Indian clothes were called generically *paños de Calicut*. I would like to remark here that the form of the name, Calicut, was also brought to Europe by the Portuguese: before Vasco da Gama, the city was called by the Italian merchants *Collicuthia* (Nicolò dei Conti), *Cholochut* (fra Mauro) and *Colocut(i)* (Colón). So the date of the *Carro de las donas*, a Spanish poem that uses the new form *Calicut*, is open to question: it must belong to the XVI, and not to the XV century, as is now commonly accepted. Calicut developped further to deformed forms like Canicul, Canicú, etc.

In 1504 all the clothes of fine linen used by Queen Isabella came from Holland (the famous *holanda*): a shirt of *holanda* was worn by the Queen on her death-bed. But Indian clothes soon rivalled the European tissues and were already exported to America in 1539. In 1542 Sancho de Moñino, the author of the *Tragicomedia de Lisandro y Roselia* (the second Celestina) enumerated, among other specimens of luxus worn by the Spanish women, the shirts of Calicut, superbly embroidered. The shirts could have sleeves or not (in the last case they were called *cuerpo de camisa*); in many occasions, the sleeves, also finely worked, were sold apart.

Indian textiles were highly appreciated also as head-dress (*tocas*). Duarte Barbosa informs us that the head-dresses, called *beatilhas* (a significant name), were exported from Diu to Arabia and Persia. Already the ships of Juan de Nova, that arrived at Lisbon in 1502, we reburdened with a cargo of *toche moresche*. In early days the head-dress was called generically *toucas de Calicut*, but soon received a more appropiate name: *bengala*, deformed in Spanish to *mengala*. The reason for the name is given by Duarte Barbosa: in the city of Bengal flourished the handicraft of woolen and cotton stuffs, greatly appreciated by the Moslems. The *bengala* was similar to the Moorish head-dress, very popular in the epoch of Queen Isabella (the most appreciated ones came from Tunis): so it received also the Moorish name *almaizar*. Sometimes it had an embellishment round the face, called in Spanish *rostriño* (*rostro* means face). The word *bengala* was registered for the first time by Dalgado and other authors in the *Comedia Eufrosina* de Vasconcelos; the Spanish documentation shows that it was introduced much earlier in Spain (where normally it is dated to the beginning of the XVIIth century), and therefore also in Portugal. In Seville a new *bengala* cost 187 mrs. in 1561; another, exported to America in the same year, was valued at 204 mrs.

Indian stuffs were employed also as bed clothes, coverlids (*colchas*), bed-sheets (*sávanas*, *covixas*), towels, etc., rivalling with the manufactures of Rouen. The set of bed clothes normally was composed of four or five cotton cloths, painted or embroidered; one of them served as cover of the bed. In 1556 a first-class set cost 5.250 mrs., but the cheaper ones oscillated between 1.600 and 3.375 mrs. The Sevillian weather, mild in winter and hot in summer, favoured the new mode. Unfortunately for our investigation, cotton bed clothes were made also in the New World, above all in Yucatán and México. So, when the documents refer only to bed clothes made in India, it is open to doubt which of the two Indias is really meant.

Among the Indian stuffs are quoted the *sinabafa*, the *canequí*, the *balagate* and the *pacharí*. The most popular without doubt was the *sinabafa*, used in shirts, head-dress, sheets, etc. In one occasion Don Quichotte said that his shoulders had been brought up in *sinabafas* and *hollands*; not so the shoulders of his shield-bearer Sancho Panza: with this comparison

Cervantes wanted to stress the social difference between master and servant. The word is registered for the first time in the Spanish poet Ercilla; but our documents show a much earlier date. The word *canequí* was introduced late in Spanish: still in 1598 it was not clearly differenciated from *sinabafa*. Cervantes spoke of long head-dresses of subtle *canequí*. As fine cloth, it was used for handkerchiefs. A servant in the comedy of Moreto *El desdén con el desdén* is called precisely *Canequí*; and a girl plays with the double meaning of the word, when she says fancifully that her nose was attracted by this *Canequí*, so that she would be pleased to have him as a handkerchief. As to the *balagate* and the *pacharí* I have only found single references to each of them. Again they are irrelevant for our purpose.

B) In the XVIth century Europe had a wide market for almost everything, but India was renowned from ancient times for its richness in gems. Precious stones appear very often in our documents. Unfortunately, their origin is normally not indicated. Only once have I noticed a mention to a ruby polished in India; and it is to be remembered that precisely a ruby was the only Indian gem Dom Manuel quoted in the letter written to Queen Isabella in 1501, announcing the return of Cabral.

Indian wood does not appear with enough frequency to draw any valid conclusion. We have only single references to boxes, tables and beds. Also a covered basket of straw appears only once.

C) The commerce of slaves began soon. They were usually born on the Malabar coast, as the skin-colour indicates: *loro*, *negro*. Generally they were very young; none of them was more than 30 years old. The first Indian slave attested in Seville is Loarte, a child of 9 bought probably in the Algarve by Vicente Yáñez Pinzón (one of the leaders of Columbus' first voyage). Their price oscillated between 7.500 and 11.250 mrs. We know that an Indian slave sent from Lisbon suffered an epileptic attack in Seville in 1515; the doctors who witnessed the attack testified before the public notary that they saw the coming of the sudden sickness: very likely the owner wanted to give back the slave to the seller, and the medical certificates were needed for that reason. The slave market had an increasing demand, but some friars denounced the terrible circumstances in which free men were enslaved without a just war intervening: so, without excuse. For that reason the owners, they said, were obliged to make a restitution, that is, to emancipate the slaves. To quiet the scruple a response was asked from Francisco de Vitoria, the famous dominican professor of Theology in Salamanca who established the criteria of just war and, therefore, the right to conquer the Indies. Vitoria replied that he could not believe that the king of Portugal would allow such a terrible commerce of slaves from his India. Again, India is a misleading term: very likely it refers here to Brazil.

Other wares do not appear frequently enough to be reckoned with in our study. However, the examples here collected show a surprisingly early arise of the "Indian mode" among the Sevillian bourgeoisie: in the dowries of the middle of the century it is customary to find an Indian head-dress or Indian bed-clothes. So, I think it is possible to speak of a certain Indianization of Spain during the XVIth century.

I. Indian Cloths

The examples are ordered

a) by the generic name

b) by the stuff

c) by the final product.

A. Generic name: "*lienços de Calicud*".

– 1 October 1520. Capitulation of friar Bartolomé de las Casas with Juan Sánchez: "he reçebido de vos... siete pieças de lienço de Calicud, a dos ducados e medio cada pieça" (937 mrs. and a half).

– 2 June 1531. Inventory of the property of the marquis of Tarifa: "Una pieça de Canicú muy delgada".

– 14 March 1539. Will of Damián de Jerez before going to the New World, declaring that he carried with him the following merchandise: "Yten confieso que llevo en mi poder a las Yndias del mar Oçéano por encomienda de Alonso de Xerez, mi hermano, quatro pieças de Olanda que tienen çiento e sesenta e ocho varas e tres quartos, e de su muger del dicho mi hermano quinze camisas de presilla labradas e dos pieças de Canicul".

– 29 November 1557. Inventory of the property of D. Pedro Núñez de Guzmán: "Una pieça de Calicud".

– 24 May 1559. Inventory of the property of Catalina Giles de Hinestrosa: "Yten un pedaço de Calicut de vara e media". Other possible Oriental goods: "iten un mondadientes de oro que pesó dos castellanos e dos tomines e medio"; "iten quatro sortijas de oro, la una con una muerte e otra con un jesú? e otra con una esmeralda e un rubí falsas e otra con un topaçio que pesaron dos castellanos e medio"; "iten unas arrecadas con sus arillos de oro que pesaron dos castellanos e dos tomines e seis granos"; "yten dos peines de marfil, digo tres peines".

– 16 August 1559. Inventory of the property of Gaspar Fernández: "Un pedaço de Canicud". Other possible object brought from Orient: "Una libra de ruibarbo"; "tres bujetas de marfil".

– 2 April 1578. Dowry of Rufina Jiménez, daughter of Pedro Trujillo and María de Luque: "Dos varas de Canicul, quatro reales".

– 15 December 1583. Auction of the property of the doctor Juan de la Fuente: "Un paño de Canicú en Graçia de Silva en seis reales e un quartillo".

B. Garments

1. *Shirts (camisas) and sleeves (mangas)*

– 20 July 1534. Auction of the property of Juan Sánchez de Robleda: "unas mangas de Calicud en Alonso Cavallero en medio ducado".

– 18 April 1540. Auction of the property of Antón de Jerez, shoemaker: "una camisa de muger con las mangas de Calicud en Alonso de Xerez en treze reales".

– 29 February 1560. Inventory of the property of Beatriz Salvago: "Yten unos cuerpos de camisa de Canicú con sus mangas".

– 2 February 1559. Inventory of the property of Mateo de Fuensalida: "Unas mangas de Canicud".

– 5 September 1584. Inventory of the property of Dª María de las Roelas, daughter of D. Pedro de las Roelas, knight of Alcántara. "Dos mangas de Canicul del tienpo antiguo blancas".

2. *Ropa*

Probably *ropa* is used *sensu lato*: "todo género de tela" (*DicAut*).

– 13 March 1558. Dowry of Isabel de Torres, wife of the bookseller Cristóbal Ramírez: "Otra ropa de Calicú blanco".

– 8 May 1558. Auction of the property of Antón Valenciano, hat-maker: "Una ropa de Canicul en Diego de Çuleta en siete reales menos quartillo".

3. *Surplice (sobrepelliz)*

– 2 January 1561. Auction of the property of Luis de Santillán: "Otra sobrepelliz de Caniqú en Gonçalo Muñoz en diez y seis reales".

4. *Head-dress (tocas)*

– 2 June 1531. Inventory of the property of the marquis of Tarifa: "Otra toca de lienzo de Canicú con los bibos y trapazejos blancos".

– 20 June 1560. Inventory and auction of the property of Francisca de la Peña: "Un pedaço de Canicud y una bolsa de carmesí vieja". *Auction*: "Un almaizal de Canicud en Juan Sánchez en siete reales y medio".

– 30 December 1584. Inventory and auction of the marquise del Valle:

- "Iten çinco tocas de Canicull por coser, xx.
- Iten quarenta y tres baras de Canicull en seis pedaços, xliij (*Auction*: se remató quarenta varas de sinabafa a real y medio la vara en Don Fernando de Pineda, que monta todo sesenta reales. Yten se remató seis baras de sinabafa, a sesenta y quatro mrs. la vara, en don Fernando de Pineda, que monta todo trezientos y ochenta y quatro mrs... Iten se remató en Diego del Almonaçir seis varas y media de sinabafa a real y medio, que monta trezientos y treinta y dos mrs... Iten se remató quatro baras de sinabafa a çinquenta y ocho mrs., que son dozientos y treinta y dos mrs. en el dicho. Yten tres pedaçillos de sinabafa en tres reales en el dicho)". As may be seen, the equivalence between *Canicul* and *sinabafa* is quite clear.
- "Una mantellina de la Yndia (*Auction*: se remató en Juan de Guarniso una mantellina blanca labrada de la Yndia en treze ducados".

5. *Bed-clothes (ropa de cama)*

The doubtful cases are grouped at the end.

– 7 October 1524. Inventory of the property of the canon D. Diego López de Cortegana: "Una cama de lienços pintados de Calicud de quatro paños"; "otro paramento de cama de Calicud"; "una antecama de lienço de Calicud"; "una pieça de pañezuelos de mesa de lienço de Calicud". Other possible objects brought from Orient (or from the Portuguese Africa): "Otra almohada labrada de seda azul con lana"; "una escrivanía de marfil"; "una tabla de marfil como portapaz con un Cruçificado pequeño"; "un mondadientes de marfil con una guarniçión de plata dorada".

– 1553. Dowry given by Ana Mallar, widow of Sancho de Herrera ("veinticuatro" of Seville): "Una cama de las Yndias de Portogal".

– 18 January 1556. Dowry of María de Medina: "Una cama de la Yndia de Portogal, que son çinco paños e sus goteras, en catorze ducados: çinco mill y dozientos y çinquenta mrs."

– 24 August 1559. Dowry of Francisca López. "Una cama de Canicud de quatro paños con una randa, en quatro ducados".

– 27 December 1559. Dowry of Beatriz de Godoy: "Una cama de Calicud con sus flocaduras en noventa reales" (about 8 ducados).

– 1 January 1560. Dowry of Leonor Ruiz: "Una cama de Canicul con sus tiras de red e sus flecaduras nuevas, e son quatro paños, doze ducados".

– 22 January 1560. Dowry of María de Baeza: "Una colcha colorada de la Yndia de Portogal, labrada, en ocho ducados... Una delantera de cama de Canicul con una tira de red, labrada, en tres ducados... Una sávana y una almohada de Canicul en doze reales... Un paño de cama labrado de la Yndia de Portogal en un ducado".

– 30 April 1560. Inventory of the property of Francisco Velázquez, schoolmaster: "Una gotera de Canicul con su flueco blanco".

– 1560. Inventory without indication of the owner (perhaps Catalina de Gámez): "Dos azerillos labrados de grana, uno de Canicud e otro de Ruán, en medio ducado. Dos paños, uno de Canicud viejos y otro de naval viejo en çinco reales. Un paño de Canicud en veinte e çinco mrs. Un paño blanco de Canicud en un real. Una xerbilleta pequeña alimanisca e un paño de Canicud viejo e roto, real e medio".

– 27 May 1560. Dowry of Mariana Ortiz: "Una cama de Yndias de algodón, seis ducados".

– 14 April 1580. Inventory of the property of Catalina de Orduña (widow of the doctor Juan Chacón), servant of Dª Juana Cortés, duchess of Alcalá: "Una colcha de Calicud.. Más unos paños colorados de la Yndia".

Doubtful Cases

– 22 November 1557. Inventory of the property of Dª Luisa de Carvajal: "Yten otra cama de Yndias de lana colorada con su cielo y quatro paños y quatro mangas, y es bieja".

– 26 August 1558. Inventory of the property of Alonso de Esquivel: "Una cama de Yndias que tiene tres paños y el çielo".

– 8 October 1558. Dowry of Francisca Marín: "Una cama blanca de Yndias de algodón en seis ducados".

– 15 November 1558. Inventory of the goods given to the hermitage of Nuestra Señora de Aguas Santas (Valverde): "Una cama de Yndias a bocadillos picada".

– 11 April 1559. Dowry of Francisca Sánchez: "Una cama de Yndias de algodón que tiene quatro paños labrada de colorado e otras colores en nueve ducados".

– 24 April 1559. Dowry of Catalina de Gadín: "Una cama blanca de Yndias que tiene çinco paños en seis ducados".

– 15 May 1559. Dowry of Ana de Vera: "Çinco paños de Yndias de algodón de çerco de cama: quatro ducados".

– 30 May 1559. Inventory of the property of Dª Antonia Portocarrero: "Una pieça de colcha de la Yndia, con cabo e cola".

– 5 August 1559. Dowry of Catalina de Estrada: "Una çercadura de cama de Yndias de tres paños y el çielo, siete ducados".

– 12 September 1559. Dowry of Constanza Farfán: "Una cama de algodón de Yndias de quatro paños, en nueve ducados".

– 18 December 1559. Dowry of Beatriz de Ribera: "Una cama de las Yndias con tiras de red labrada, en treze ducados".

– 25 August 1561. Dowry of Dª Leonor Cataño Carranza: "Una colcha de seda de Yndias colorada en doze ducados".

– 7 September 1561. Dowry of Violante Cabrona: "Una cama de Yndias que tiene quatro paños pintados, quatro ducados".

– 18 October 1560. Inventory of the property of Inés de las Casas: "Una colcha de Yndias".

– 11 March 1562. Inventory of the property of the lute-maker Sebastián Rodríguez: "Una cama de paños de Yndias con tres colchones de lana que tienen cuatro arrobas de lana, con seis sávanas y seis almohadas blancas y un cobertor de Valensia blanco con sus tablas y sus bancos, que vale todo CCL reales".

– 24 April 1562. Will of Dª Catalina de Villavicencio. Miguel de la Cruz has "una cama de paños de Yndias, que tiene tres paños", that belongs to her daughter Elvira.

– 1596. Inventory of the property of Pedro de la Torre Espinosa, banker. "Yten una covixa labrada de la Yndia de Portugal". He also had "un Anus Dei de évano guarneçido con plata".

– 1596. Inventory of the property of the doctor Simón de Tovar: "Una colcha mediana de la Yndia de Portugal".

6. *Towels (Toallas or hazalejas)*

– 27 May 1538. Inventory of the property of Elvira de Góngora: "dos azalejicas de Calicud labradas de blanco".

– 17 September 1559. Inventory of the property of Diego de la Peña, chief of police: "Unas toallas de Canicul"; "unas tovajas de Canicul con un flueco"; "una almohada de red. Otra de Canicul".

– 7 July 1560. Auction: "Yten un pedaço de Canicul como paño de rostro en Diego de Flores en tres reales (102 mrs.)".

7. *Plaited cloths (trenzados)*

– 29 May 1559. Auction of the property of Dª Catalina Riquelme: "tres trençados de Canicul en treinta y ocho mrs."

C. Stuff

1. *Bengala/Mengala* (*port.* *bengala*)

– 19 February 1557. Dowry of María Hernández: "Una toca mengala e una gorguera guarnesçida de oro en un ducado".

– 17 March 1557. Dowry of Juana Pérez: "Dos bengalas y una toca de algodón".

– 15 May 1558. Dowry of María Rodríguez: "Dos mingalas y otras cosas de tocados en quinientos e sesenta y dos mrs."

– 15 April 1559. Dowry of Isabel Gallego: "Una mengala con su rostriño".

– 30 September 1559. Goods given to Eufrasia Pérez: "Una mengala".

– 26 June 1560. Inventory of the property of Elvira Núñez, wife of Bartolomé Bernardo, gold-embroiderer: "Una toca mengala de muger traída".

– 12 October 1560. Inventory of the property of the surgeon Gabriel de Colunga: "Una toca mingala".

– 7 April 1561. Dowry of Isabel de Cárdenas: "Una mengala nueva, medio ducado" (187 mrs.).

– 10 April 1561. Merchandise given by the shoemaker Francisco Jiménez to the ship-master Alonso Fernández: "Yten una bengala que costó seis reales" (204 mrs.).

– 9 December 1561. Dowry of María Sánchez: "Un almaizal e un paño de cabeça e una mengala e unas mangas de cotonia e un jubón de muger, un ducado".

– 2 May 1562. Dowry of Catalina López: "Quatro varas de mengala, un ducado".

– 1 April 1578. Inventory of the property of Simón Díaz: "Treinta varas de mingala de algodón".

– 2 April 1578. Dowry of Rufina Jiménez: "Dos pieças de mengalas, que tienen treinta varas, a setenta e dos mrs. cada una".

– 1 August 1578. Dowry of Isabel de Reinoso: "Dos mingalas finas e dos tocas de seda en dos ducados e medio".

– 7 May 1582. Inventory of the property of Juana de Pineda: "Más una mingala toca".

– 8 August 1583. Inventory of the property of Leonor Rodríguez: "De sinco varas y media de mingalas de seda a sinco reales y medio la vara, montan trenta reales y ocho mrs."

2. *Balagate* (*port.* *balagate*)

– 2 April 1578. Dowry of Rufina Jiménez: "Veinte e siete varas de balagate, a real e medio cada vara".

3. *Canequí* (*port.* *canequim*, *1546 [Dalgado, I, p. 202 b]*)

– 30 October 1597. Inventory of the property of the canon Pedro Fernández de Castro: "Una pieça de canequí. Dos paños de Indias paa (paes? para?) sobrecamados frontales, uno de red y otro de un tafetancillo".

– 12 September 1598. Inventory of the property of Juan de Medina: "Una colcha de sinabafa o caniquí nueva".

4. *Pacharí*

– 4 March 1538. Diego López, refiner of the mint, acknowledges a debt of 20.000 mrs. to Pedro de Castro for 50 "pieças de lienços pacharíes de la Yndia de Canicul".

5. *Sinabafa* (*port.* *sinabafo*, *1508 [Dalgado, II, p. 306 a]*)

– 3 March 1555. Dowry of Jerónima de Polanco: "Una camisola de sinabafa blanca con sus mangas de tafetán encarnado con sus bueltas de franjas de oro: ocho ducados".

– 15 April 1559. Gregorio Ortiz buys from Leonor de la Vega some merchandise, among other things "quatro gorgueras de sinabafa... çinco reales".

– 17 September 1559. Inventory of the property of Diego de la Peña: "un pedaço de sinabafa nueva en que ay doze varas escasas".

– 22 October 1559. Dowry of Leonor García: "Una cama que tiene quatro paños de lienço de sinabafa con sus flocaduras en doze ducados".

– 3 November 1559. Execution in the property of Hernán Gómez Adalid: "Tres cuellos de sinabafa".

– 9 November 1559. Dowry of Francisca López: "Otro paño de manos de senabafa, labrado de verde, dos ducados".

– 1560. Inventory of the property of the canon Constantino de la Fuente. "seis varas de lienço sinabafa".

– 8 January 1560. Dowry of Ursula López: "Más un paño de rostro blanco y un frutero y otro paño de sinabafa blanco en un ducado".

– 28 February 1560. Division of the inheritance of Ginés Farfán, husband of Dª Catalina Núñez: "una saya colchada de sinabafa en çinco ducados".

– 10 April 1561. Merchandise given by the shoemaker Francisco Jiménez to the ship-master Alonso Fernández, who intended to sail to Santo Domingo touching at Cabo Verde: "Dos gorgueras de sinabafa que costaron ocho reales".

– 6 April 1562. Inventory of the property of the licentiate Juan de Quevedo: "Una ropilla de sinabafa traída".

– 9 January 1578. Auction of the property of the priest Pedro Vélez: "Otra sobrepelliz de çinavafa en Luis Sánchez, clérigo, en doze reales".

– 29 July 1578. Inventory of the property of Alonso de Ávila: "Una saya de sinabafa frisada vieja".

– 24 September 1579. Inventory of the property of Francisco de Ávila: "Una toca de camino de sinabafa vieja".

– 4 April 1582. Will of María de Esquivel: "Mando a Leonor Ximénez... una ropilla blanca de sinabafa".

– 14 April 1582. Dowry of Dª Francisca de Esquivel Ponce de León: "Una colcha de sinavafa grande en trezientos reales".

– 16 April 1582. Dowry of Inés Pérez: "Una colcha de entretela de çinabafa quatro ducados".

– 5 September 1584. Inventory of the property of Dª María de las Roelas: "Yten unos manteles de sinabafa".

– 30 December 1584. Inventory and auction of the property of the marquise del Valle. "Varias varas de sinabafa".

II. Wood

Painted wood

a) ***Tables (mesas)***

– 22 November 1557. Inventory of the property of Dª Luisa de Carvajal: "Yten dos mesas con sus bancos, la una de Yndias pintada y la otra de nogal". Doubtful.

– 1598. Inventory of the property of Diego Alemán: "Una mesa pintada de la Yndia con su banco de cadena".

c) *Bed frame (armadura de cama)*

– 21 June 1556. Dowry of Catalina de Santa Cruz: "Una armadura de cama de madera pintada de colores de la Yndia de Portugal en diez ducados".

– 19 April 1581. Donation of fray Antonio de Zúñiga, brother of the marquise del Valle, to Lucas de Medina: "Una cama de madera amarilla y negra de la Yndia, con sus cinchas".

III. Slaves

– 10 February 1513. Vicente Yáñez Pinzón sells to the canon Alfonso de Escobar a slave, "loro"-coloured, called Loarte, about 9 years old, "natural de Calicud", por 9.000 mrs.

– 2 April 1515. Pedro Fernández and Fernando de Jerez sent slaves to his father Diego de Jerez from Lisbon. Among them was Pedro, born in Calicut, whom the licentiate Antonio de Cuenca, doctor, saw suffering an attack of the disease that "entre médicos se llama epilepsia y en romançe se dize gota coral" at Diego de Jerez' house; the sick slave "no tornó en su acuerdo desde las nueve oras antes de mediodía fasta la noche", according to the testimony of the merchant Bartolomé de Jerez.

– 11 December 1531. Juan de Lucio, pharmacist, sells to Juan Gutiérrez de los Ríos "un esclavo de color loro, de Canicul, que ha nonbre Françisco" for 21 ducados (7.875 mrs.).

– 4 May 1532. The Burgalese merchant Francisco Yáñez de Burgos sells to Gutierre del Castillo "un esclavo de color loro, indio de Canicul, que a nonbre Gaspar, de hedad de veinte e ocho años poco más o menos", for 9.000 mrs.

– 21 January 1538. Pedro Fernández Franco sells to Julián de Tudela an Indian slave, called Juana, 18 or 20 years old, "con unas letras en el carrillo derecho que dizen *Pedro, natural de la Yndia de Portugal*" for 50 ducados (18.750 mrs.).

– 19 March 1539. Diego Gómez, inhabitant of Cáceres, sells to the wine-merchant Diego de Saavedra "un mi esclavo de color loro, indio de la Yndia de Canicul, que ha nonbre Antonio, de hedad de diez e siete años poco más o menos", por 20 ducados (7.500 mrs.).

– 25 April 1539. Francisco García, inhabitant of Lisboa, sells to Juan de Alvarado and to María de Camargo, his wife, "un esclavo de color negro, indio de Canicud, qu'es del rey de Portugal, que a nonbre Christóval, de hedad de diez y siete o diez e ocho años", for 30 ducados (11.250 mrs.).

Doubful Cases

– 25 May 1536. Fernando de Aldana sells to the tailor Juan Fernández, "un esclavo negro de Portugal", Juan, 24 years old, for 5.000 mrs.

– 17 February 1540. Antón Sánchez, town-crier, sells to Gonzalo Martín Cisneros an "esclavo indio de la Yndia de Portugal, que a nonbre Juan Merín", 10 years old, for 9.000 mrs.

– 1 April 1540. Francisco Fernández, slave-merchant, sells to Martín Barrasa an "esclavo indio, natural de la Yndia de Portugal, que ha nonbre Françisco, de hedad de veinte años poco más o menos", for 30 ducados (11.250 mrs.).

– 2 August 1553. Francisca Rodríguez, widow of the leather-dresser Juan de la Puebla, sells to the leather-dresser Miguel Jerónimo "un mi esclavo indio de la Yndia de Portugal, que se diz Sebastián", for 34 ducados (12.750 mrs.).

– 26 November 1555. The brothers Jerónimo y Juan Rodríguez, potters, sell to the esparto-worker Juan Librero "un nuestro esclavo indio de la Yndia de Portugal, que a nonbre Pedro, mediano de cuerpo, con barbas, de hedad de treinta años poco más o menos;... el qual dicho esclavo está herrado en la cara en anbos carrillos con unas letras que dizen: 'Alonso de Herrera, vezino de Puerto Real'", for 12.000 mrs.

– 1 February 1559. Gómez de Prado sells to Pedro de Campo "un esclavo indio de la Yndia de Portugal que a nonbre Pedro, de hedad de veinte e çinco años poco más o menos", for 9.000 mrs.

– 1 July 1559. Alonso de Villanueva sells to the shoemaker Francisco Gutiérrez an "esclavo de color loro, indio de la Yndia de Portugal, que a nonbre Martín, de hedad de treinta años", for 40 ducados (15.000 mrs.).

– 24 April 1560. Juan Luis de Ribera, inhabitant of Seville, sells to Bernardo de Ribas "un mi esclavo de color loro natural de la Yndia de Portugal, de hedad de veinte e seis años poco más o menos", for 60 ducados (22.500 mrs.).

– 28 November 1560. Bartolomé de Tovar, inhabitant of Zafra, sells to Ruy López de Ribera "una esclava india de la Yndia de Portugal que a nonbre Ynés, de hedad de veintiquatro años", for 66 ducados (24.750 mrs.).

– 12 May 1561. Account of the merchant Luis Alonso. "Reçebí más del dicho jurado Hernand Pérez una esclava por nonbre Polonia, india de las de la Yndia de Portogal".

– 25 May 1561. Juan Vanegas sells to Juan Cabado an "esclavo indio de la Yndia de Portugal, barvudo, herrado en la cara, que a nonbre Diego, de hedad de treinta años poco más o menos" for 27 ducados (10.125 mrs.).

– 22 August 1596. Francisco de la Barrera, inhabitant of Cádiz, has an slave "de color mulato, indio de las Indias de Portugal, que se llama Françisco, de hedad de treinta años".

JESUIT OBSERVATIONS AND STAR-MAPPINGS IN BEIJING AS THE TRANSMISSION OF SCIENTIFIC KNOWLEDGE

HASHIMOTO Keizo
Kansai University

In this paper, I will discuss the transmission of astronomical knowledge to China, particularly the methods of the star mapping, during the Jesuits' existence in Beijing, considering the importance of the role played by some of the eminent missionaries, i.e., Matthew Ricci, Adam Schall von Bell, Ferdinand Verbiest, and Ignatius Kögler. Some of our investigations show that the scientific missionaries not only transmitted the very recent knowledge at their times from Europe, but also they actually observed heavenly bodies, and created new knowledge, for example, through the compilation of the introduced astronomical methods for the would-be calendrical reform in the end of the Ming dynasty.

1. Christopher Clavius in Rome and Matthew Ricci in Beijing

In the beginning, I should like to show that Matthew Ricci (1552-1610), using the later edition, i.e., the edition of the year 1593 (*nunc quarto*), translated Clavius's *Sphaera* into Chinese in Beijing. The edition can be specified, because the illustration of the structure of the cosmos shows the 11th sphere (***Fig. 1-a***[1] and ***1-b***). This means that Christopher Clavius (1537-1612), following Copernicus's arrangement, finally refuted the theory of trepidation in the edition of that year.[2] We can confirm this fact by examining the Chinese version, translated by him, that is, the *Qiankun tiyi* 乾坤體義 (Beijing, 1605).[3] Here we can also show the title page of Clavius's *Sphaera*

---

[1] Cf. James M. Lattis, *Between Copernicus and Galileo*, The Univ. of Chicago Press, 1994: p.39.
[2] *Ibid.* p.164.
[3] *Quan* 1, 9a-b.

of 1596 against the illustration in the Chinese version, so as to compare them to each other (***Fig. 2-a***[4] and ***2-b***).

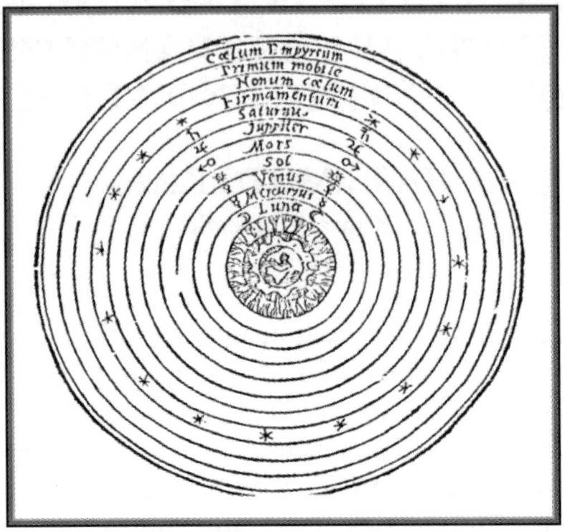

Fig. 1-a – Clavius's depiction of the eleven-sphere cosmos. From Clavius's *Sphaera* (Venice, 1596; James M. Latis, *Between Copernicus and Galileo*, The University of Chicago Press, 1994, p.39).

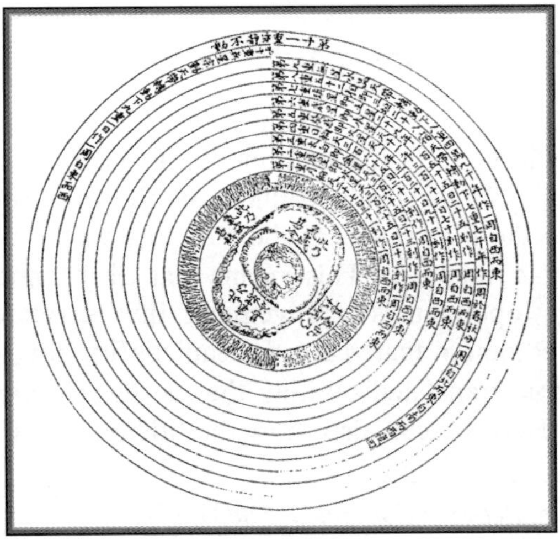

Fig. 1-b – The Chinese version of Clavius's depiction of the eleven-sphere cosmos in the *Qiankun tiyi*.

[4] Fig. 2-a is from Lattis, *op. cit.*, p.40.

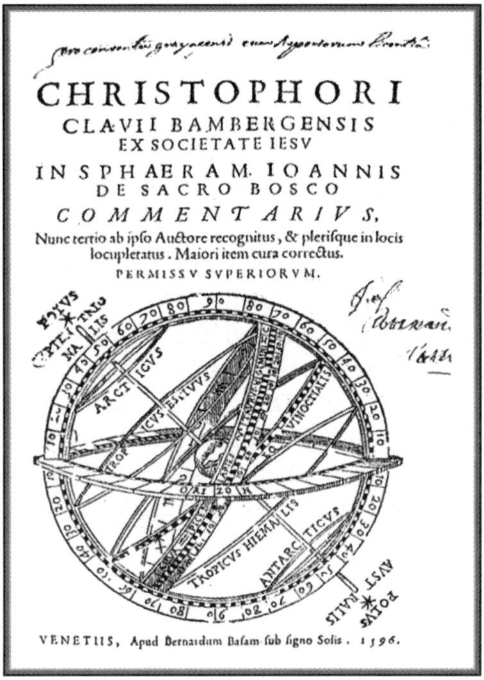

CHRISTOPHORI
CLAVII BAMBERGENSIS
EX SOCIETATE IESV
IN SPHAERAM IOANNIS
DE SACRO BOSCO
COMMENTARIVS,
Nunc tertio ab ipso Auctore recognitus, & plerisque in locis locupletatus. Maiori item cura correctus.
PERMISSV SVPERIORVM.

VENETIIS, Apud Bernardum Basam sub signo Solis. 1596.

Fig. 2-a – The Title page of Clavius's *Sphaera* (Venice, 1596; Latis, *op.cit.* p.40).

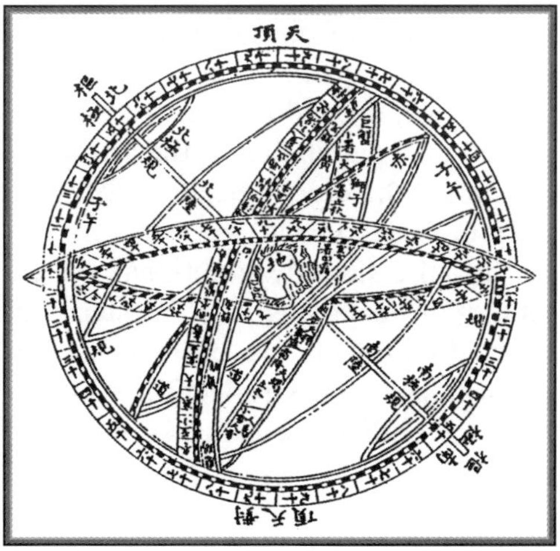

Fig. 2-b – The Chinese version of the *sphaera materialis* from the Qiankun tiyi.

This also means that Ricci, Clavius's old student, was using the newest edition available to him for arranging the Chinese version. The translation was important for the later missionaries as the basis of the general introduction of Western astronomy. And the introduced natural philosophy of this sort eventually influenced Chinese intellectuals, including the author of the *Wuli shaozhi* 物理小識, Fang Yizhi 方以智, Xiong Mingou 熊明遇, and so on, in Late Ming and Early Qing periods.

One of the scientific missionaries, who belonged to the second generation of the missionaries, Johann Adam Schall von Bell, SJ (1591-1666), Tang Rowang 湯若望, later compiled the *Treatise on the Gnomon*, the Chinese title of which was the *Rigui tufa*. This was the translation of the *Gnomonices libri octo*, in 1581, by Clavius. The Qing copy of the manuscript of this has been preserved at Kyoto University.[5] I would like to point out this fact, because Adam Schall, too, was one of the last students of Clavius at the Roman College, so as eventually to translate his teacher's treatise in China.

2. Star-Maps prepared in Beijing in Late Ming China

First of all, I should like to make clear of the contribution of Johann Adam Schall von Bell during the compilation of the *Chongzhen lishu* 崇禎曆書 under the leadership of Xu Guangqi 徐光啓 (1562-1633) from 1629 concerning the present topic. I should like to single out his case because he, together with Xu, opened up the new stage of star-mappings in China, which symbolized the compilation of the astronomical encyclopaedia at that time. It also occupied the crucial position for the transmission of the scientific knowledge from West to East.

Since China had had the strongly traditional system of star-catalogue and the unique method of mappings, the Western knowledge was not capable of replacing it. As a result, we observe a kind of the reconciled achievements in the star-atlases. They reflect the Chinese style of representation of stars in spite of with more exact knowledge and observational results from the West.

---

[5] The Chinese name of Johann Adam Schall von Bell has been transcribed not as 湯若望, but as 湯如望 for the author of the book.

Adam Schall manufactured the four kinds of Star-Maps:

1) *Jianjie zongxingtu* 見界總星圖,

2) *Qidao nanbei liang-zongxingtu* 赤道南北兩總星圖,

3) *Huangdao nanbei liang-zongxingtu* 黃道南北兩總星圖, and

4) *Huangdao ershifenxingtu* 黃道二十分星圖.

In addition, two kinds of them, 1) and 2), were specially mounted and separately presented to the throne under the different titles from the list of presentation, i.e., the *Hengxing zongtu* 恆星總圖 (The General Star Atlas) in 1631 and the *Hengxing bingzhang* 恆星屏障 (The Screen of Star-Map) in 1633, respectively. And the third star map was also independently manufactured.

We find the aim and intention of the compilation of the *Chongzhen lishu* declared by Xu materialized in both of them particularly in the General Star Atlas 恆星總圖. We also find the symbolic meaning of the repeated presentations of such star-maps to the thrones by Jesuit missionaries in between late Ming and early Qing dynasties. When Adam Schall and Xu originally compiled the *Hengxing lizhi* 恆星曆指 (Treatise on the Theories of Fixed Stars), they intended to prepare the four kinds of illustrated star-maps as the *quan* 4 of it, but, they eventually separated this part of the treatise as independent, and gave the title of the *Hengxing jingwei tushuo* 恆星經緯圖説 (The Illustrated Treatise on the Positions of Fixed Stars).

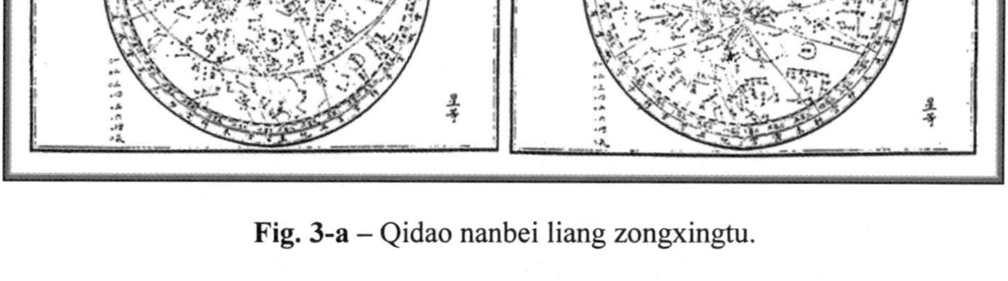

Fig. 3-a – Qidao nanbei liang zongxingtu.

It is interesting to find that Adam Schall originally relied on the *Catalogus* (Rome, 1612), published by the successor of Christopher Clavius and his old teacher at the Collegio Romano, i.e., Christopher Grienberger. But, eventually, Adam Schall must have increased the considerable number of stars observed in Beijing. Here we also have to point out the strong influence observed in the star maps, particularly in that of the *Qidao nanbei liang-zongxingtu* 赤道南北兩總星圖 (The General Star Atlas of the Northern and Southern Equatorial Double Hemispheres, ***Fig. 3-a***). At the same time we must not overlook the symbolic meaning materialized, among others, in the *Jianjie zongxingtu* 見界總星圖, the point of which we have made clear in our previous work.[6]

3. Christopher Grienberger and the *Catalogus* (Rome, 1612)

Christopher Grienberger was the successor of Clavius at the Collegio Romano as the professor of mathematics. According to Galileo, he was regarded to belong to the same group with himself.[7] His *Catalogus veteres affixarum Longitudines, ac Latitudines conferens cum novis* was published in Rome in 1612, the same year when Clavius died. This is the star catalogue prepared at the College. The stars catalogued have been taken mainly from Tycho Brahe's *Astronomiae instauratae progymnasmata* in 1600, in which the star number amounts to 767, but the substantial number of stars were supplemented by the observations by Clavius as well as Francesco Pifferi, 240 and 218, each.

These stars are classified according to the northern and southern hemispheres. The total number amounts to 1225 stars, consisting of 670 in the northern and 555 in the southern sky, respectively, have been listed in the equatorial coordinates, the right ascension and declination, respectively, in the second part of the book. They have all been transformed from the original ecliptic coordinates, the positions of which had also been given in the first part of the book, because it was traditional in Europe from Greek times. According to Joseph Needham, this fact itself shows the very important and characteristic change in the development of the history of astronomy in general.[8]

---

[6] K. Hashimoto, 「見界總星圖と『恒星總圖』」, 京都大學人文科學研究所研究報告『中國古代科學史論　續編』, Kyoto Univ., 1991-3: 333-366.
[7] Rivka Feldhay, *Galileo and the Church*, Cambridge UP., 1995: pp.240-241. D'Elia suggests that Grienberger was a Copernican too. Cf. Lattis, 1995: p.205.
[8] J. Needham, *The Grand Titration*, Allen & Unwin, 1969: pp.78-80.

Grienberger also prepared the detailed images of constellations, consisting of 26 pieces of parts of the sphere in the equatorial coordinates, which can be observed in his *Catalogus*. And, the main chart included in the book is the Star Atlas of the Double Equatorial Hemispheres (***Fig. 3-b***).

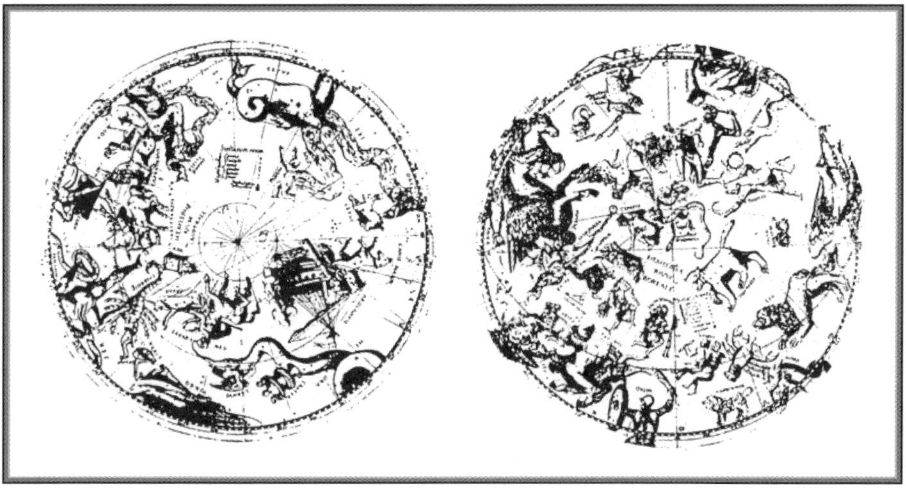

Fig. 3-b – Equatorial Star Map (C. Grienberger, *Catalogus*, Rome, 1612)

We can say that the Star Atlas, altogether, gave him the essential information when Adam Schall was trying to prepare the *Qidao nanbei liangzongxingtu* (***Fig. 3-a***). And the set of 26 pieces of the constellation charts of Grienberger seems to have given the crucial clue when he was preparing the *Huangdao Ershi-fenxingtu* (Star Chart of the Twenty Divisions in the Ecliptic Coordinates, ***Fig. 4***). According to the *Hengxingbiao* (the Star Catalogue), which has been included in the *Chongzhen lishu*, the star positions were adjusted from those of Grienberger's *Catalogus* by making use of the table of precession included in the *Catalogus* as well (***Fig. 5-a***, ***b***, ***c***, and ***d***). The epoch is the year of 1600 while that of Adam Schall's Chinese version was the year of 1628, that is, the first year of the Chongzhen reign-period.

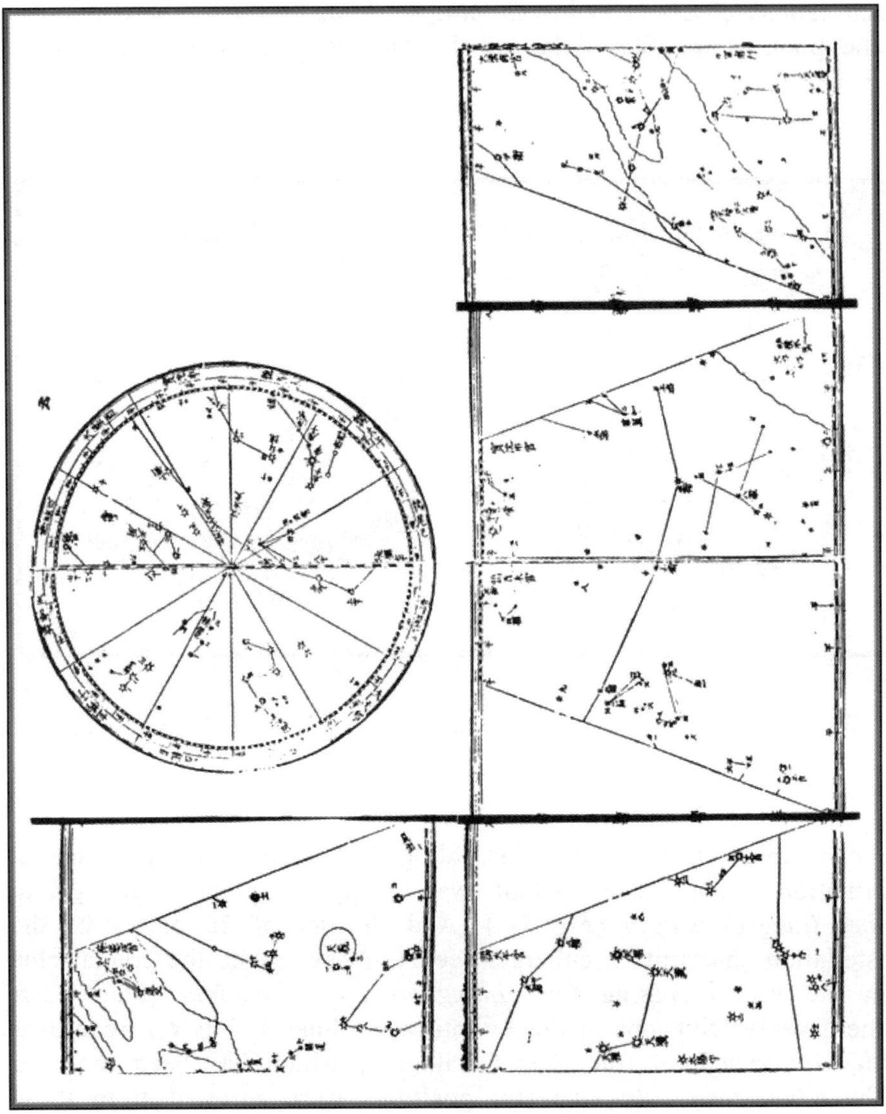

Fig. 4 – Huangdao ershifen xingtu.

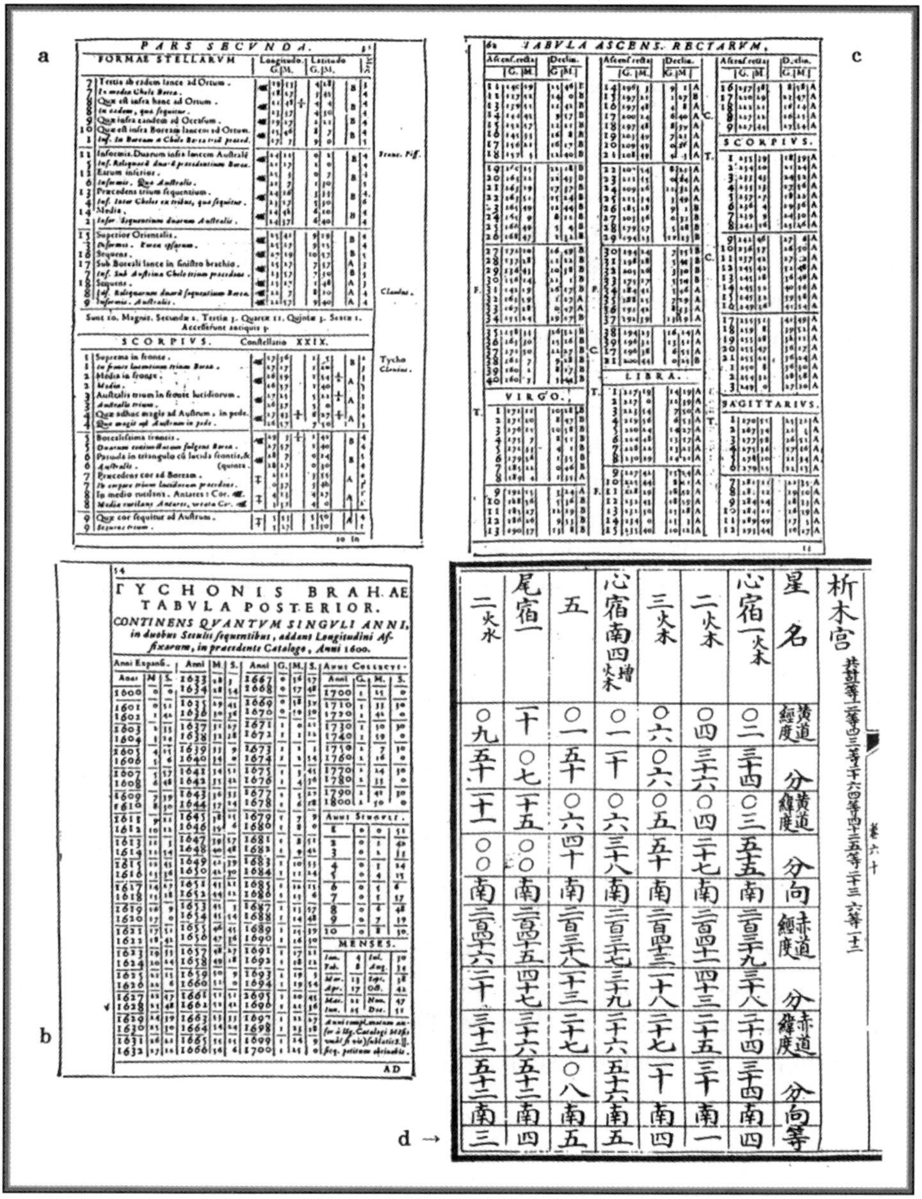

Fig. 5 – The comparison of the *Hengxing biao* in the *Chongzhen lishu* (5-d) with Grienberger's *Catalogus* (5-a, -b, and -c).

At the beginning of the astronomical reform, Xu Guangqi had proposed to manufacture several astronomical instruments to observe the positions of heavenly bodies. As early as 1630, John Schreck (Terrenz) had one set of Large Sextant manufactured for the observations of fixed stars before his death late in that year. He also had two sets of the Large Quadrants manufactured for the observations of the seven luminaries.[9]

The Sextant must have been used for the position observation of the fixed stars. The difference of the star numbers between Grienberger's *Catalogus* and Chinese *Star Catalogue* in the *Chongzhen lishu* provides one of the evidences of the fact of this matter.

The *Qidao nanbei liangzongxingtu* was eventually enlarged into eight pieces of sheets, which was altogether mounted as a large screen, so as to present to the throne as the *Hengxing bingzhang* (Screen of the Fixed Star) as the part of the achievement of the reform (***Fig. 6***). This star atlas altogether has a large size. The height is about 170 cm, and the width 400 cm. The fact that the screen was mounted from 8 pieces of sheet altogether can be confirmed by examining the Paris version of the atlas.[10]

According to the *Hengxingbiao*, the number of stars listed in this Chinese Star Catalogue is 1351, which completely outnumber the stars listed in Grienberger's *Catalogus*. The number eventually increased to 1812 according to the description of the explanation of the independent *Qidao nanbei liangzongxingtu*. The observation of those stars, which completely outnumbered the original *Catalogue*, was obviously carried out by the astronomers in Beijing, including Adam Schall von Bell, during the reform.[11]

We have found the Chinese constellation system was not switched to the European system at all. This shows the traditional Chinese system had been too well established to be changed. The Jesuit astronomers had to satisfy to conform it to the Chinese system.

---

[9] Cf. Liang Jiamian, *Xu Guangqi Nianpu* (A Chronicle of Xu Guangqi), Shanghai, 1981: p.176.

[10] K. Hashimoto, 「赤道南北總星圖と『恒星屏障』」, 京都大學人文科學研究所『新發現科學史資料の研究』, Kyoto Univ., 1985-11:581-604.

[11] See note 9.

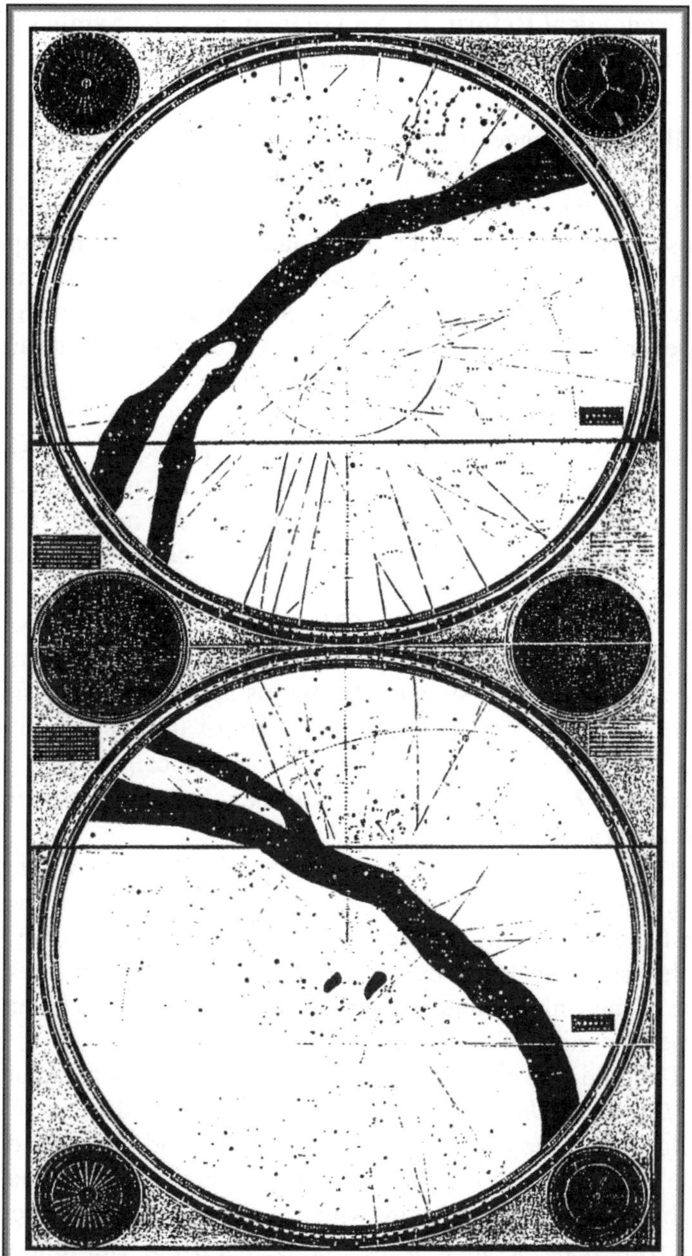

Fig. 6 – Qidao nanbei liangzongxingtu, or Hengxing bingzhang.

4. The Astronomical Reform by Xu Guangqi and the Symbolic Meaning of the *Jianjie Zongxingtu* Planisphere

The illustration page of the *Jianjie zongxingtu* (Planisphere, or the General Star Atlas of the Whole Visible Sky) alone are missing from the extant version of the *Hengxing jingwei tushuo* (Treatise on the Theories of Stellar Positions). And, yet, the explanation of this star-map itself is included in the first pages of the treatise. This is rather mysterious only if we just pay our attention to the fact of the missing of this characteristic star-map. However, I was rather lucky to discover this separated sheet of the star-map at the Bibliothèque Nationale de Paris at the end of the year 1982 (***Fig. 7***).

According to the *Book Catalogue* of the Bibliothèque,[12] the sheet of the star map is named the *Hengxing jingwei tushuo*. I was eventually able to make clear of what happened to the part of the star map of this treatise. The folios of this map had, as a matter of fact, been separated from the original binding of the treatise at a very early stage.

In the second presentation in 1631, this Star Atlas was mounted and eventually presented to the throne as an independent item, carrying the title, the *Jianjie zongxingtu*. The style of the map is apparently quite similar with the *Suzhou Star Map* in 1247. The planisphere of the representation of the visible sky with the North Pole in the centre is called the *Gaitu* 蓋圖 (Planisphere, see ***Fig. 8***). And, this newly manufactured planisphere looks like very much traditional.

But, at the same time, the representation of the apparent magnitude of stars, among others things, is quite European. We should not overlook the geometrical method of the projection of the visible sky, either. In many respects, this star-map was the symbolic achievement of the result of the confluence of the Chinese and Western traditions, which Xu Guangqi really aimed at the beginning of the enterprise of compilation.

Here I should like to show how the original star map included in the *Hengxing jingwei tushuo* looks like, and how is the difference from the *Jianjie zongxingtu*. The outer diameter is 57.4 cm, and the inner diameter is 55 cm. The size of the folio 10-a and -b attached to the sheet of the Star Map is the height 27.7cm x width 34cm, which is the same as with the book size of the *Hengxing lizhi* included in the *Chongzhen lishu*. The folio carries the explanation of the last part of the fourth Star Map, *Huangdao ershifengtu*.

[12] Maurice Courant, *Bibliothèque Nationale Catalogue des Chinois, etc*, Paris, 1910.

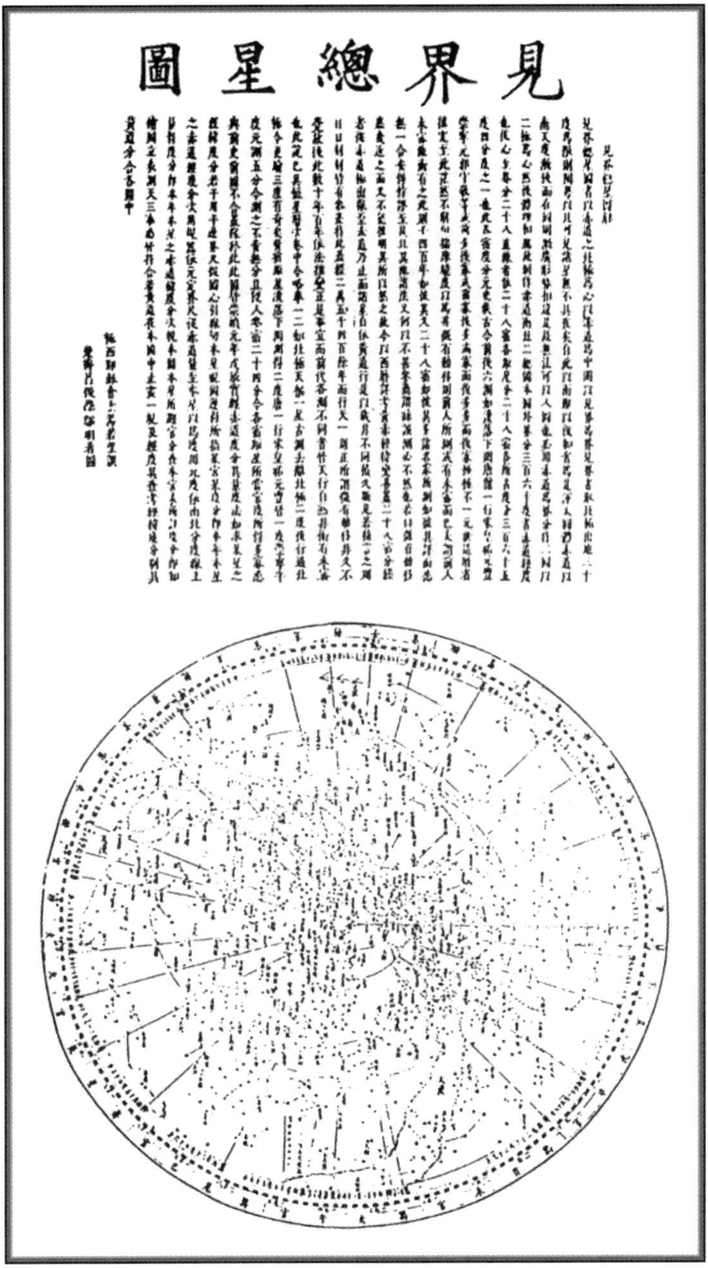

Fig. 7 – Jianjie zongxingtu in the *Hengxing jingwei tushuo* at the Bibliothèque nationale in Paris.

This is the evidence that these parts were altogether separated from the original treatise, *Hengxing jingwei tushuo*, as we have just mentioned above.

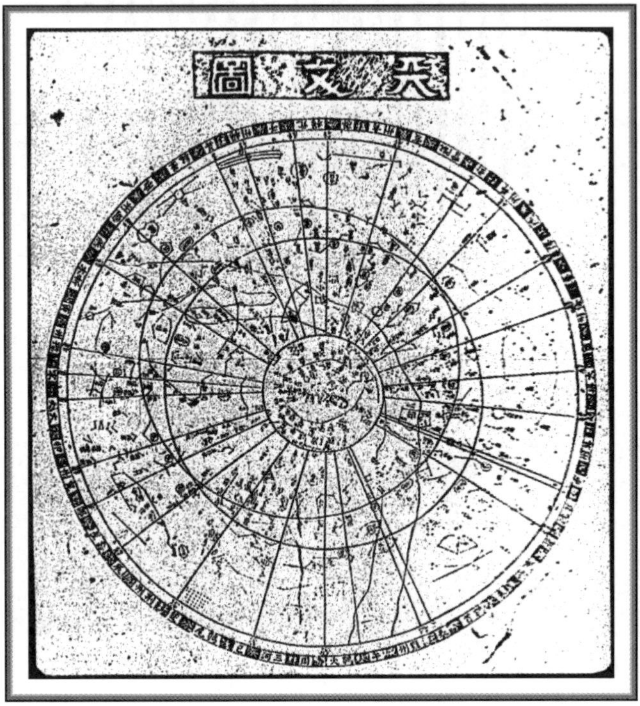

Fig. 8 – Suzhou Star Map (1247).

The representation of the separately mounted independent Star Atlas, *Jianjie zongxingtu*, is slightly different from this original Map. The outline of the Milky Way, for example, is drawn double dotted. This is an obvious difference between the original and the independent star map. The Astronomical Chapter of the *Ming shi* writes that Xu Guangqi presented the *Jianjie zongxingtu* to the throne in the beginning of the Chongzhen reign-period.[13]

[13] It was probably in 1631. Liang Jiamian, *Xu Guangqi Nianpu*, Shanghai, 1981: p.204. The *Ming shi* also writes that, later in 1633, the *Qidao nanbei liangzongxingtu* was presented as well.

5. The Influence of the Recent European Observations with the Telescope on Star-Atlases in the Qing Period

We also should not forget that the results of the developing telescopic observations were immediately introduced into China and influenced the star mappings in China. It is rather surprising if we see the precise illustrations that reflected such development in Europe. I would like to show you one example from the Star Maps in the early eighteenth century.

According to the star atlases manufactured by Ignatius Kögler (Dai Jinxian 戴進賢, 1680-1746), the recent results of the telescopic observations in Europe were vividly reflected on the map. We can show this fact in the *Huangdao zongxingtu* 黄道總星圖 (General Star Atlas of the Ecliptic Double Hemispheres), which we examined by making use of the version preserved at the Royal Bibliothèque of Albert II in Brussels (***Fig. 9***), which we happened to find there. This marks the new stage of the introduction of European astronomy in the eighteenth century. The atlas has been dated in 1723.[14] He did not mind to represent the celestial by the equatorial coordinates.

The sun is shown with the sunspots on the surface. The moon's surface is characteristically represented with the results of the developed telescopic observations. What is noted is the representation of Saturn. The ring is shown with the very recent discovery of the five satellites. According to the explanation, F.-B. Moggi (Li Baiming 利白明, 1684-1761) has scripted the map. They were based on the recent observational results obtained at the Paris Observatory. The illustration of the moon seems likely to have been adopted from the *Astronomia reformata* (1665) by G.-B. Riccioli in Bologna. The text eventually became the crucial basis for Koegler to adopt Kepler's Laws without changing the world system of Tycho Brahe. The problem will be discussed separately later on.

Apart from the constellation system and the presentation of the 24 seasonal divisions on the circumferencial scale, the map can be said rather Western. This fact reflects that Kögler might not have accustomed to the Chinese system yet at the still earlier stage of his stay there. Eventually he systematically observed the whole sky to produce the star catalogue, called *Yixiang kaozheng* 儀象考成, eventually dedicated to the Emperor Qianlong in 1746.

---

[14] The first year of the Yongzheng reign-period, 雍正元年.

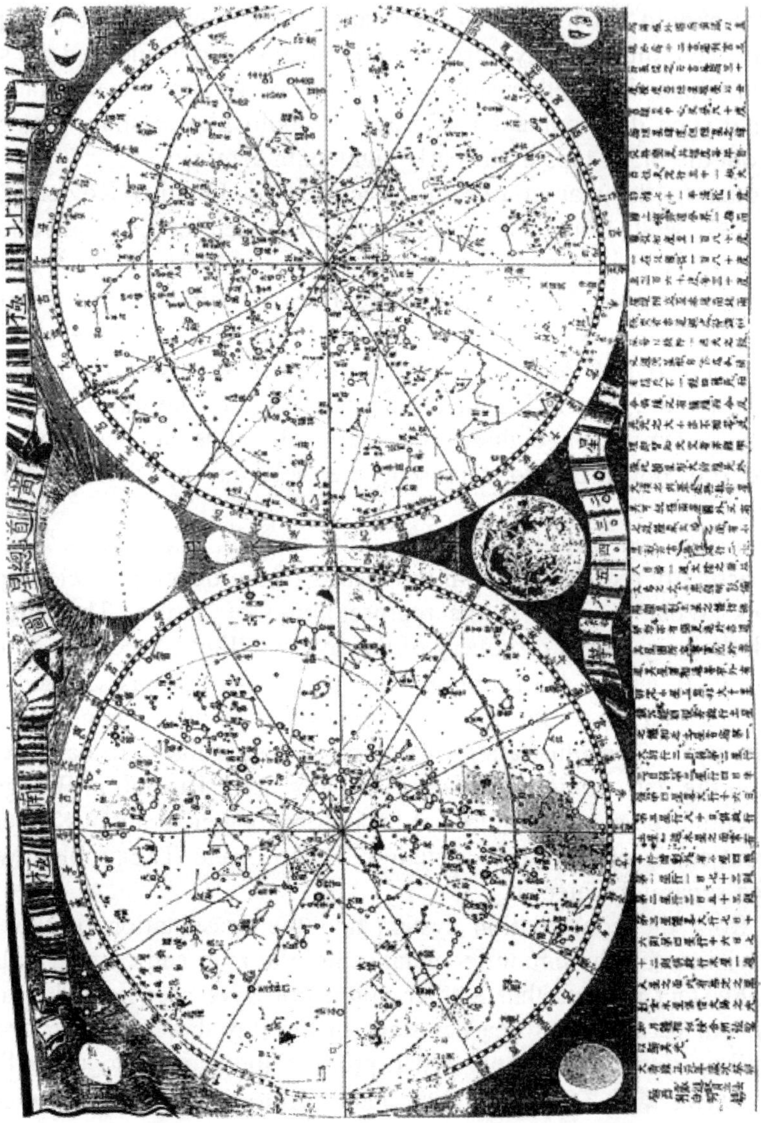

Fig. 9 – Huangdao zongxingtu preserved at the Royal Bibliothèque of Albert II in Brussels.

It is well known that he, together with Andreas Pereira, took the responsibility to translate into Chinese the *Luni-solar Tables*[15] of P. Grammatici, who had used the data reduced from Newton's achievements, so as officially to introduce Kepler's Laws into China in 1742.[16] It was the completion of the *Lixiang kaicheng hou bian* 曆象考成後編. But, we do not discuss this matter any more.

Concluding Remarks

We have examined the transmission of the knowledge of star mapping as a result of Jesuit arrival in Beijing. As a whole, it is very impressive that, although the coordinates system and the methods of projection were introduced to replace the traditional system, the constellation system was not switched to the Western system. We can say that, concerning what happened, the cultural aspect of astronomy was too strongly rooted in the tradition to be changed to the introduced system. We must remember that this sort of way to introduce the new, Western knowledge was what Xu Guangqi meant at the beginning of the compilation of the astronomical compendium, based on European astronomy.

[15] Joseph Dehergne, SJ., *Répertoire des Jésuites de Chine de 1552 à 1800*, Rome and Paris, 1973; p.137.

[16] See Han Qi's paper in this book.

THE COMPILATION OF THE *LIXIANG KAOCHENGHOUBIAN*, ITS ORIGIN, SOURCES AND SOCIAL CONTEXT

HAN Qi

Institute for the History of Natural Science, Chinese Academy of Sciences, Beijing

In recent years historians of Chinese science have studied in great depth the introduction of Western science in China. They mainly worked on the transmission of European mathematical sciences during the Chongzhen and Kangxi reign.[1] However, no detailed research has been done on the spread of European astronomy in the Yongzheng-Qianlong period.[2] Based on some new European and Chinese sources, I will analyse the social context of the introduction of Newton's lunar theory into China in 1734.

In 1993, I had an opportunity to do some research at the Bibliothèque de l'Observatoire de Paris, where I found a Chinese astronomical book, entitled *Lixiang kaocheng biao* (Tables of the Compendium of Astronomy). [3] As far as I know this is the only printed copy extant in the world. With this table we can find on a piece of paper a few lines that read as follows:

> *"Tables Lunisolaires du R. P. Grammatici, traduites en Chinois et imprimées à la Chine, par les soins des R.R.P.P. Ignace Koegler et André Pereira qui me l'ont envoié avec leur lettre commune du juillet 1734, ils ont ajouté les titires des tables ecrits en Latin sur des bandes de papier inserées dans le livre.* [...]

[1] See Hashimoto Keizo, *Hsu Kuang-ch'i and Astronomical Reform.* Osaka: Kansai University Press, 1988. Joseph Needham, *Science and Civilisation in China*, vol.3: *Mathematics and the Sciences of the Heavens and the Earth.* Cambridge: Cambridge University Press, 1959.

[2] On Copernicus in China, see Nathan Sivin, "Copernicus in China", *Studia Copernicana*, 6 (1973), pp.63-122.

[3] Library of the Observatory of Paris, A B1/11.

L'on voit par la lettre du P. Koegler au P. Grammatici du 15 Juillet 1732 (dont j'ay pris copie) que ces tables etoient alors deja traduites en Chinois; mais pas encore imprimées; L'on y trouve aussi quelques particularitez sur ces Table qu'il faut voir au lieu cité."

This autograph was written by Joseph-Nicolas Delisle (1688-1768), a French astronomer and geographer. As seen above, we know that the lunisolar tables were translated into Chinese and printed in China, based on the original tables of Nicaise Grammatici (ca.1684-1736). They were sent to Delisle in July 1734 by the Jesuit astronomers Ignatius Koegler (1680-1746) and André Pereira (1689-1743), who were directors at the imperial board of astronomy (Qintianjian, QTJ) at that time.

First, we will talk about very briefly about Grammatici. He was born in Trente, Italy. From 1701, he devoted his life to Christianity. As a Jesuit priest, he was also a very good astronomer. In 1722-1726, he was a professor of mathematics and Hebrew at the College of Ingolstadt in Germany. He made a lot of astronomical observations in Madrid (1727-1728) and in Trente (1729). When he was in Spain, the Spanish King Phlippe V ordered him to teach mathematics at the College of Nobles. Later he returned to Germany and died in Regensburg in September 1736. He published several astronomical books, one of which was entitled *Tabulae lunares ex theoria et mensuris Geometrae celeberrimi Domini Isaaci Newtoni Equitis Aurati in gratiam Astronomiae Cultorum concinnatae a quodam Uranophilo e Societate Jesu* (Ingolstadii, 1726, or *Lunar Tables according to the theory and measurements of the mathematician, the most famous Mr. Isaac Newton...*).[4] From this title, we know that it was based on Newton's lunar theory.[5] In this book, there were three pages of explanation, twelve diagrams on eight pages. According to his preface, Grammatici not only used Newton's books, but also cited the books of G.D. Cassini, J. Cassini, E. Menfredius and D.E. de Louville.

Second, I would like talk about Delisle's life. Born in Paris, he learned astronomy from Maraldi, a royal astronomer at the Observatoire de Paris. Later he became a famous astronomer and geographer and was nominated as a professor at the Royal College in 1718.

---

[4] On the scientific works of Grammatici, see Carlos Sommervogel, *Bibliothèque de la Companie de Jésus*. Bruxelles-Paris, 1890 sq.

[5] This is the first lunisolar table which was based on Newton's theory. In their letters, the Jesuits of China mission also stressed that it was based on Newton's mechanics. See Han Qi, "The discovery of the first Chinese translation of Newton's *Principia*", in *Zhongguo keji shiliao*, vol.19, no.2 (1998), pp.78-85.

In the early 18th century, a number of eminent French scientists were invited to work in Russia by Peter the Great. In 1725, Delisle arrived in St Petersbourg in order to establish an observatory and stayed there until 1741. He made a lot of astronomical observations. He was also a good collector, bought and copied a large number of astronomical observations made by many famous astronomers. He not only had much correspondence with the European scientists, but also kept up close contacts with the Jesuits in Beijing when he was in St Petersbourg.[6] Through Russian diplomats and the caravans, he received from the Jesuits many astronomical observations and some Chinese books which were compiled by them. The lunisolar table (*Lixiang kaocheng biao*) was one of them.

During the Yongzheng reign the astronomers at QTJ could not predict the heavenly phenomena accurately according to the theory of the *Lixiang kaocheng* (Compendium of Astronomy). Therefore the Jesuits, with Ignatius Koegler and others in charge, compiled the *Lixiang kaocheng houbian* (A supplement to the *Lixiang kaocheng*), using many new European astronomical achievements. Koegler was from Bavaria (Germany), arrived in Beijing in 1717 and served for the emperor. In 1725, he was promoted to the director of QTJ. He stayed there for twenty-nine years and played a leading role in the compilation of the *Lixiang kaocheng houbian*.[7]

As mentioned above, the Chinese lunisolar tables were translated from Grammatici's book. However, when and how did Grammatici's book arrive in China?

As early as in 1728, Koegler received Grammatici's lunisolar tables. The French Jesuit Antoine Gaubil (1689-1759) wrote to his confrère:

> *"J'ai bien trouvé le moyen d'avoir ici quelque chose d'Angleterre et d'Allemagne en manuscrit, savoir des Nouvelles tables lunisolaires sur le système de M. Newton, des tables des satellites tirées des méthodes de M. Pound, et bien des observations faittes à Nuremberg, à Ingolstad et ailleurs. J'avois déja des éphémérides, mais médiocres, du marquis Ghisler, tout cela m'étoit venu les années*

[6] See Knud Lundbaek, *T. S. Bayer, Pioneer Sinologist.* London: Curzon Press, 1986. When Delisle was in Paris, he met a Chinese Christian from Fujian, named Arcade Hoange (1679-1716). It was probably through Hoange that Delisle began to be interested in Chinese science, especially the history of Chinese chronology and astronomy.

[7] On Koegler, see Louis Pfister, *Notices biographiques et bibliographique, sur les Jésuites de l'Ancienne mission de Chine, 1552-1773*, Chang-hai. 1932. Han Qi, "Biography of I. Koegler", in Du Shi-ran (ed.), *Zhongguo gudai kexuejia zhuanji* (Xiaji), Beijing: Kexue chubanshe, 1993, pp.1330-1332.

passées par le P. Koegler qui le tenoit du P. Grammatici, mathématicien d'Ingolstad." [8]

That is to say, Grammatici sent his book directly to Koegler. Because Koegler also stayed at Ingolstad for a period of time, they had close contact with each other. Therefore Grammatici sent Koegler many new astronomical publications in Europe and also tried to publish his astronomical observations.

After having received the lunisolar tables, Koegler began to change them into Chinese form. But it was quite a job for him. With the help of André Pereira, the vice-director of QTJ, he spent at least four years before its publication in about 1734. Shortly after its publication, Koegler and Pereira sent it to Delisle. In order that Delisle could understand the Chinese lunisolar tables, they added the Latin names of tables on the relative pages.

One may ask: who played the most important role in the compilation? Why did Koegler choose Grammatici's book as a basis for Chinese tables?

Before we move on to the context of the compilation of the *Lixiang kaocheng houbian*, we will say something about the compilation of the *Lixiang kaocheng*. In 1711, when the Kangxi emperor realised that the length of the sun's shadow at the summer solstice that he observed was not in accordance with that calculated at QTJ, he decided to learn mathematical astronomy from the Jesuit missionaries. In 1713, he issued an imperial decree to open the *Mengyangzhai* (Studio for the Cultivation of the Youth) to compile and translate the *Lixiang kaocheng* and the *Shuli jingyun* (A mathematical encyclopaedia).[9] In 1723, the *Lixiangkaocheng* was printed at the imperial printing house.

However, because the *Lixiang kaocheng* still used Tycho's astronomical system in the calculation, it could not meet the needs of precise prediction, especially in predicting solar and lunar eclipses. As a director of QTJ, Koegler was responsible for these predictions. As soon as he got Grammatici's new lunisolar tables, he realised that they were much more precise than those tables of the *Lixiang kaocheng*. This was the reason why he chose it as the basis for Chinese tables.

---

[8] See Antoine Gaubil's letter to E. Souciet, dated 26 November 1728, in *Correspondance de Pékin*. Geneve, 1970, p.219.

[9] On the social context of the compilation of the mathematical and astronomical books, see HAN Qi, "Emperor, Prince and Literati: Role of the Princes in the Organization of Scientific Activities in Early Qing Period", in Yung Sik Kim & Francesca Bray ed., *Current Perspectives in the History of Science in East Asia*, Seoul: Seoul National University, 1999, pp.209-216.

As mentioned above, Delisle kept very close contact with the Jesuits in Beijing. Through his correspondence with the European astronomers, even with Grammatici, he knew of developments in European astronomy. In his letter to the Jesuits in Beijing, he proposed a scientific exchange program and asked them to send him their astronomical observations. On the other hand, he promised to tell them the new astronomical discoveries in Europe. Therefore as soon as he got any new information, he kept them informed. On 20 September 1731, he wrote a letter to the Jesuits in Beijing, which reads as follows:

> *"Since I wrote you last time, I have not heard from any new progress in mathematics and astronomy. Only in England someone was improving lunar theory based on Mr. Newton's theory. An able mathematician from the Royal Society of London, named Machin discovered some direct and more easier methods to solve most of the problems by which Newton got the principles of inequality of the moon. [...] Therefore, my dear Fathers, I would not forget to keep you informed about that."* [10]

Apparently, Delisle was very keen on the new development of the lunar theory. This also aroused the Jesuits' interests in Newton's lunar theory. This can also explain why the lunar theory based on Newton's principles was introduced into China in about 1732.

We have mentioned Koegler's important role in the compilation of the *Lixiang kaocheng biao*. In fact, he found that the *Lixiang kaocheng* was not precise even earlier than Chinese astronomers. He was engaged in the calculation, but he had not reported to the emperor. However, according to the Chinese official document, it was not until 1730 that QTJ found the observation of solar eclipse was not in accordance with the prediction. At this point Mingtu, the director of QTJ, sent a memorial to the Yongzheng emperor, asking the revision of the *Lixiang kaocheng*. This plan was approved by the emperor. In the same year, the lunisolar tables were completed as a supplement to the *Lixiang kaocheng*. Because there was no explanation of the method for calculation, only Koegler and Pereira and the Mongolian astronomer Ming Antu could use it. For this reason, in 1737, Gu Cong, the minister of Rites, presented a memorial to the Qianlong emperor, in which he suggested that digrams and explanations should be

---

[10] Library of the Observatory of Paris, v.115 XVI. Corr. Delisle 1731-1734, T.IV. n.23.

added. He recommended Mei Juecheng (1681-1763) and He Guozong (?-1766) to be general editors. This is why the *Lixiang kaocheng houbian* was compiled. Koegler, Pereira, Mei Juecheng and He Guozong were in charge of the calculations.[11]

In the eighteenth century, the astronomers of the Observatory of Paris had kept close contact with the Jesuits in China. Many of the astronomical translations were from the works of French astronomers. But for the sake of precision, the applications of the results of the Jesuit scientists were often used for astrological purposes at the court. In late imperial China, astronomy still played an important role at the imperial court. To predict very precisely the solar and lunar eclipses was especially important. If people found that the prediction was not accurate, the officials at QTJ would be punished. For this reason, Koegler and his Chinese colleagues made attempts to compile the lunisolar tables and this led to the compilation of the *Lixiang kaocheng houbian*. After the lunisolar tables were compiled, they met the needs of precise prediction. This is the reason why Newton's lunar theory was indirectly introduced into China no later than 1734 and why the *Lixiang kaocheng houbian* was used in China for quite a long period of time. [12]

Acknowledgements: I am grateful for the grant (19803006) from the Natural Science Foundation of China in supporting this research.

---

[11] Concerning the Jesuit astronomers' role at QTJ, see HAN Qi, "The Role of the Directorate of Astronomy in the Catholic Mission during the Qing Period", in N. Golvers ed., *The Christian Mission in China in the Verbiest Era: Some Aspects of the Missionary Approach*. Leuven: Leuven University Press, 1999, pp.85-95.

[12] In addition to the lunisolar tables, J. Richer's observation of Mars at Cayenne, Kepler's equation, the theories of procession, parallax, refraction, and the obliquity of ecliptic were introduced in the *Lixiang kaocheng houbian.*

A JAPANESE REACTION TO ARISTOTELIAN COSMOLOGY

YOSHIDA Tadashi
Centre for Northeast Asian Studies, Tohoku University

The British historian Charles Boxer once called the years 1549-1650 "the Christian century" in Japan.[1] The Christian influence has tended to be underestimated because of the successive persecutions and executions against Christians by the authorities towards the end of the sixteenth century and around the beginning of the following century, as well as the *sakoku* policy which was introduced in 1630s. It is true that the paucity of written documents on the part of Japanese authors gives us tremendous difficulty in trying to trace historical facts for this period. However, there remain a few sources which tell us what was brought in by the missionaries.

Having noticed the inquisitive and rational mind of the Japanese, Francisco Xavier (1506-52) and Alessandro Valignano (1538-1606), Visitor to Japan and China, both of whom had laid the foundation for missionary activities in Japan, adopted a policy of teaching them first the order and beauty of the natural world and then inviting them to recognize the existence of God behind it. Valignano also found it necessary to teach Japanese Christians the subjects in the humanities and his idea led to the establishment of the college (Colegio de Sao Paulo) in Funai in 1580. It started with a group of seven or eight young foreigners who were taught Latin grammar and the basics of the Japanese language and humanities studies. They completed these studies in 1583 and moved on to the full course in philosophy. When Pedro Gomez (1533-1600), a Spanish Jesuit, reached Japan in 1583, one of his most urgent tasks was the organization of studies at the Funai college. Gomez was asked to write a textbook for the college

---

[1] C.Boxer, *The Christian Century in Japan (1549-1650)* (Berkeley/Los Angeles: University of California Press, 1951)

students on account of his previous experience of teaching at the Jesuit College in Coimbra. Sometime between August and October in 1593, his long-awaited textbook was completed. By 1593, more than forty students, Europeans as well as Japanese, are said to have been studying at the college. Since Japanese students knew no Latin, Pedro Ramon (1550-1611), another Spanish Jesuit, translated the textbook into Japanese probably by the beginning of 1595. This textbook, usually called *Compendium* which is the title of the third part: *Compendium catholicae veritas...*, consists of three parts:

(1) *de sphaera* in which Aristotelian cosmology and meteorology are discussed,
(2) *de anima* in which Aristotelian philosophy on soul is explained, and
(3) the theological parts in which Christian faith and post-Tridentine Roman Catholic theology are treated.

Thus, Gomez's trilogy became the standard textbook for Japan. It is the first part of his trilogy, *de sphaera*, that played a significant role in the introduction of Aristotelian cosmology into Japan.[2]

1. Sawano Chuan and the *Kenkon Bensetsu*

Christovao Ferreira (1580-1650) from Lisboa had been engaged in missionary activities in the Kyushu area since the early 1610s before he was caught in Nagasaki in 1633. Unable to endure the torture of his captors, he became an apostate and naturalized, taking up a new name Sawano Chuan. He was ordered by Inoue Masashige, Inspector General against the Pagans, to translate into Japanese a Western astronomy book, which was confiscated from a Christian father caught in 1643. Chuan (Ferreira) wrote the text of translation in Roman alphabets, because he could not freely write in Japanese scripts, though he was well versed in Japanese. A certain interpreter named Nishi Kichibei converted the text of Roman alphabets into Japanese *kana* and *kanji*. Then, Mukai Gensho (1609-77), a Confucian scholar, made

[2] For the transcription and new Japanese translation of Gomez's text of *de sphaera*, as well as the introduction to the background of Gomez and his treatise on the sphere, see Obara Satoru, "Kirishitan jidai no kagaku shiso" (Scientific thought in Christian century), in *Kirishitan Kenkyu*, vol.10, (1965).

critical comments on the Western cosmological ideakanj put forward by Chuan. This is how *Kenkon Bensetsu* (Discussions on the Heavens and the Earth with Critical Comments, preface 1650) came into being.[3] The Western book on astronomy mentioned above which Chuan is said to have translated has not yet been identified. But the textual comparison reveals that Gomez's *de sphaera* is certainly one of the main sources for his translation. However, some of the illustrations in the *Kenkon Bensetsu* can not be found in Gomez's textbook, but can be seen in Christopher Clavius' commentary on Sacrobosco's *de sphaera* (1607). Clavius' commentary deals only with the heavenly world and not with four elements and meteorological phenomena in the sublunary world which the *Kenkon Bensetsu* discusses. This fact reveals that Chuan consulted more than a couple of Western astronomy books, and the sources for his translation seem to be not so simple as told in Mukai's preface.

2. Comparison of the Characteristics of Learning among Four Countries

Mukai Gensho writes a short treatise at the beginning and before the start of Chuan's texts in order to compare the characteristics of learning among four countries: Japan, China, India and the West. He first notes that Japan, whose small islands are scattered around in remote places, is able to be called a great country because of people's wit, virtue, and high morality. Then he divides the four countries into the two groups: Japan-China and India-the Southern Barbarians (the West). Its demarcation principles are the following two:

(1) writing system; vertical or horizontal and

(2) eating style, using chopsticks or eating directly with the hands.

[3] This is based on the preface by Mukai Gensho. However, his preface is dated wrongly, and doubts have been cast on its authenticity. Chuan's preface is dated in 1650. One of the different versions *Nanban Unkiron*, has a note that the hand-written copy was acquired in Kyushu during the Shimabara campaign (1637-8). If it is true, the *Kenkon Bensetsu* must have been completed by this time. However, this is contradictory to the passage in Mukai's preface, where he says that Chuan died without giving any title although his text had already been completed. Mukai's remark suggests that the *Kenkon Bensetsu* was completed not long before Chuan's death in 1650. Although the date of completion is yet to be decided, I think that the actual process of how Ferreira, Nishi, and Mukai were involved is not so different from what has been told in the preface. See Preface, *Kenkon Bensetsu,* newly printed in *Bunmei Genryu* Sosho, vol. 2 (Kokusho Kankokai, 1914), pp. 1-2.

The countries of the first category, users of vertical writing and chopsticks, to which Japan and China belong, adopt the concepts of *li-qi, yin-yang*, and five elements as the main principles of their learning, while the countries of the second category, horizontal writers and users of hands in eating, to which India and the Western countries belong, do not know the theories of *li-qi*, and *yin-yang*, and believe in the four-element theory, casting doubt about the concept of five elements. The learning of the Southern Barbarians are well versed in the theories which concern figures[4] and instruments[5] and do not have deep knowledge about things which go beyond figures (i.e. metaphysical meanings). The countries in the first category are, in his opinion, obviously superior to those in the second category. Both in Japan and China, Confucianism is believed and their learning is based on the theories of *li-qi*, and *yin-yang*. However, Mukai argues, the basis of Japanese learning is Shintoism. Without it, even Confucianism will degenerate into the learning which seeks only for utility. Students of Shintoism who are not well versed in Confucianism shall fall into heresy and witchcraft. Therefore, to master both Shintoism and Confucianism is the best way for learning.[6] In Mukai's opinion, Japan, thus, comes first, with China coming after her in the standard of learning.[7]

3. Aristotelian Four Elements vs. Confucian Five Elements

The *Kenkon Bensetsu* consists of four parts, two on the heavenly world (*ken*) and two on the sublunary world (*kon*), following Aristotelian dichotomy between them. In contrast to the order of Gomez's text of the *de*

---

[4] *Katachi* or *kei* in the original text is translated as "figure" throughout this paper, because I am afraid that "form" may be misconceived as something to do with Aristotelean *forma*.

[5] Though translated as "instrument", *ki* in *keiki* in the original text comes from the usage contrasted to *Tao* in the *Book of Change*, where Tao is posited at metaphysical level, while *ki* is at the physical level. Therefore, Mukai's intention here is to point out that learning in India and the West is good at the physical level, but not at the metaphysical level.

[6] *Kenkon Bensetsu,* pp. 6-8.

[7] For a general discussion on the *Kenkon Bensetsu* in English, see Nakayama Shigeru, *A History of Japanese Astronomy* (Cambridge, Mass.: Harvard University Press, 1969), pp.88-98, and Mikami Yoshio, "On an astronomical treatise ("*Kenkon Bensetsu*") composed (in 1650) by Portuguese in Japan", *Nieuw Archief voor Wiskunde*, vol. 10 (1913), 61 ff.

sphaera, the *Kenkon Bensetsu* starts with discussions on the sublunary world, where the first seven chapters are devoted to the introduction of the concept of four elements. In his commentaries on the text of the *Kenkon Bensetsu* prepared by Ferreira, Mukai criticizes Western ideas mostly from the viewpoint of Confucian *yin-yang* and five-element theories. One of its typical examples is the confrontation between Aristotelian concepts of the four elements and Confucian theory of the five elements.

Chuan's first criticism is targeted at the idea that myriad things are made out of the combination of the four elements. Following Aristotelian dichotomy, the heavens have nothing to do with the generation and degeneration of matter in the sublunary world. Mukai's dissatisfaction with the four elements lies precisely on this point. It is told, he points out, in the very beginning that the light of the sun, the moon and stars illuminate the world and nurture myriad things on the earth, raising them by showering virtue all over them. In the beginning of the part on the celestial world, it is said that myriad things on the earth are generated out of a combination of the four elements, but there is nothing which generates without receiving the vital *qi* from the Heavens. Therefore, he argues that the scholar of the Southern Barbarians is contradictory in his discussion. Mukai does not seem to be able to comprehend Aristotelian dichotomy to its full extent, for he counts the heavens as one element of the traditional five-element doctrine, saying that

> "[...] *since the five elements are supplied by the function of* yin-yang, *which is obvious as we see things right in front of us, the barbarian scholar cannot conceal one element. Therefore, he discusses the heavens in addition to the four elements.* [...] *Is it true that the heavens plus the four elements make the five elements? Is it not the case that the combination of the five elements generates myriad things?"* [8]

Chuan (Ferreira) attempted to explain the four-element theory in terms of the five-element principle by employing the concepts of production (*sosho*) and overcoming (*sokoku*). For instance, he explains, since "dry" in fire and the one in earth are of the same *qi*, fire "produces" earth, although "warm" in fire and "cold" in earth are contradictory to each other. But, "dry" in earth and "humid" in water are of different *qi* so that earth overcomes water.

[8] *Kenkon Bensetsu*, p. 54.

Such misuse of Chuan as to the concept of production and overcoming is severely criticized:

> *"The concept of production and overcoming is beyond the comprehension of the students of the South Barbarians".*

Being the head of the Confucian school in Nagasaki, Mukai could not tolerate Chuan's application of his superficial knowledge of Chinese philosophy. His conclusion is: the reason why they argue like this is they do not know the theory in its true sense.[9]

After Chuan's discussion on the Aristotelian concept of the natural place of four elements where earth holds the lowest place, Mukai argues in his refutation that when we place water on the earth, water easily goes down and submerges in the earth, while when we place earth on water, earth will be sunk into water. Therefore, he concludes, the truth is that earth and water hold the same place, both being mingled together and not be able to be separated.[10] Mukai here is no doubt confused about elements as a theoretical concept with earth and water in the real world.

4. The Sphericity of the Earth

Strongly influenced by the Chinese traditional idea (*gai tien* theory) of the "round heaven and square earth" and by the Sumer cosmology of Buddhism, the Japanese had adopted the theory of the flat earth. The idea of the sphericity of the earth was brought by Catholic missionaries in the "Christian century". For instance, Luis Frois (1532-97) reports that Oda Nobunaga (1534-82) had a terrestrial globe at his side which had been presented by them. A famous debate took place between a Confucian scholar and a Japanese Christian in Kyoto in 1606. One of the subjects over which they disagreed was the shape of the earth. Hayashi Razan (1583-1657) who was to become the founder of the *bakufu* school for Confucianism wrote a short essay about this debate under the title of the *Hai Yaso* (Rejecting the Christians)[11]. The man who vindicated Christian ideas was called Fukan Fabian (1565-1621?), a man of strange fate, who joined in the Society of Jesus in 1586 and wrote the book *Myotei Mondo* (Dialogue between Two

---

[9] *Kenkon Bensetsu*, pp. 9-16.

[10] *Kenkon Bensetsu*, pp. 20-23.

[11] The text for the *Hai Yaso* can be found in *Nihon Shiso Taikei* vol. 25 (Kirishitan sho, Haiya sho [Christian and anti-Christian books]) (Iwanami Shoten: 1970).

Nuns) in which he explained Christian concepts, but became an apostate soon after the debate and attacked Christianity, in turn, in the book *Ha-Daiusu* (Refuting Deus).

Razan was first shown by the picture of Jesus and the map of the spherical earth. They started the discussion on the concept of the direction "up" and "down". According to the theory of the flat earth, the direction "up" always refers to the direction over our heads while "down" to the direction beneath our feet, wherever we are. However, in the theory of the spherical earth, the direction beneath one's feet ("down") points to the direction over the heads ("up") for the man situated at the opposite side of the round earth. The treads of the two men who stand precisely at the opposite side of the earth face towards each other with the earth between them. This phenomenon was incomprehensible and the whole discussion was ridiculous for Razan. Shown optical instruments like prisms, he made a comment that the Christians devised curious instruments by which to allow the innocent populace go astray. At one point, Razan asked Fabian the question "who created the Creator, then?" After a series of exchanges of their views, Razan concluded: Christians claimed that both the heavens and the earth were round, whereas, in his judgement, there were motion and rest, and things were round and square; everything was in accordance with these principles; and thus, the heavens and the earth were not the exceptions. Following the traditional Chinese concept, Razan drew the principle: those that move are round, those that rest are square. This obviously shows that the square and flat earth rests while the round heaven moves.

Given this situation, Mukai's comments on the sphericity of the earth in chapter 8 of the *Kenkon Bensetsu* is quite remarkable. It was probably the first time that this idea was explicitly recognized and approved. Chuan enumerates various pieces of evidence for the sphericity of the earth:

(1) the time for dawn and sunset is different according to the location along the longitude (the sun would rise and set at the same time everywhere if we adopt the theory of the flat earth);

(2) when we are leaving the land on a ship, we can see a port and seashore while we are not so far from the land, but can not see them from far away;

(3) when we are approaching the land, we first see the tops of mountains and are only gradually able to observe the lower parts.

The cases (2) and (3), Chuan argues, can happen because of the curvature of the earth.

(4) At the time of the Great Discovery, the Spanish set sail westwards, while the Portuguese went eastwards, and they often encountered each other somewhere near China or Japan. (If the earth is flat, this will never happen. Two ships sailing in opposite directions will never meet.)

(5) The shape of the shadow (of the earth) at the time of lunar eclipse is round.

Although he affirms the sphericity of the earth, the tone of Mukai's comments sounds apologetic. Pointing out that the Confucians already noted the idea by an analogy that the heavens and the earth were spherical like the shell and yolk of an egg or a bullet, he argues that the Confucians do not claim that the earth is square in shape, when they talk of the idea of "round heavens, square earth", but that "square" in this phrase simply refers to the four cardinal points. He admits that "although the theories proposed by the Southern Barbarians are not erroneous, quoting clear examples and evidences, their discussions remain merely on the level of figures and bodies, and are moreover tedious". Since those who are well versed in the *yin-yang* theory and the principles of the heavens and the earth, will easily understand the principle (of the sphericity of the earth) without listening to the evidences brought up by the barbarians, and the idea of the sphericity of the earth was already discussed in the *hun tien* theory (the analogy of egg's yolk above), the idea that the heavens and the earth are spherical should be learned in Confucian teachings.[12]

5. Physical vs. Metaphysical

In sharp contrast to the first two parts which deal with the sublunary world, the latter two parts of the *Kenkon Bensetsu* on the subject of the heavenly world receive less severe criticisms from Mukai. His general comments are made in such a mild tone as "we do not find any errors,"[13] or "no objection".[14] He goes as far as to say with such praise as "the least erroneous. Very good arguments" or "the idea of the barbarians is supreme".[15]

---

[12] *Kenkon Bensetsu*, pp. 25-29. The idea of the "square" earth was often defended by the argument that it did not refer to the shape but to the "virtue" (*toku*). See, for instance, *Nigi Ryakusetsu,* part 2, chapter "the area for the element earth is spherical" in *Nihon Shiso Taikei* vol. 63 (Iwanami Shoten, 1971), p. 98.

[13] *Kenkon Bensetsu*, pp. 58, 70, 71, 74, 84, 87, 92, and 96.

[14] *Kenkon Bensetsu*, pp. 74 and 77.

[15] *Kenkon Bensetsu*, pp. 64 and 85, respectively.

However, throughout the whole book we repeatedly encounter his remarks that the barbarians do not know the principle of *li-qi*, or that of *yin-yang*. As we noted above (Section 2), the criteria for his evaluation of the learning in the four countries are exactly this *li-qi* and *yin-yang* theories.

However, we should note that Mukai recognizes Western superiority at the level of figures and instruments *(keiki)*, i.e. in physical matters. Let us quote a typical argument of his:

> *(after a brief explanation of Aristotelian four-element theory) "[...] thereby they have doubts about the theory of five elements and do not know the theory of* yin-yang. *They are good only in elaboration on the level of figures and instruments (*keiki*) and therefore ignorant of metaphysical (*keijijo*) meanings. Although they doubt, they do not dare to discuss (metaphysical meanings). Their discussions are detailed only on the subjects regarding figures (*katachi*). Therefore, they are not good in insights about the past, present and future, and easily go astray in such a nonsense as paradise and hell. They do not know the principles of* li-qi *and* yin-yang *even when they talk about the subjects over figures and thus fall into banal as well as vulgar discussions".* [16]

The terms *keijijo* (metaphysics) or *ki* (instruments) in *keiki* appear in the *Book of Change*, where *ki* is contrasted to Tao. Therefore, we can interpret the words like figures (*katachi*) or instruments as something physical. What Mukai intended to say in the above quotation is that the Westerners are good at physical investigations whereas they are virtually ignorant of metaphysical discussions. Needless to say, by metaphysical discussions he meant those which were argued on the basis of the principles of *li-qi* and *yin-yang*. He repeats in his comments that "they do not know *yin-yang*" or "they are not versed in the theory of five elements".[17] Moreover, Mukai notes that when the barbarians elaborate their discussions merely on the level of figures,[18] they appeal to examples and evidences,[19] which appears tedious. Mukai seems to have been impressed by the Western way of argument, but this, in turn, led him to believe that the Westerners were good only at superficial (physical) discussions and not at metaphysical arguments.

---

[16] *Kenkon Bensetsu*, p. 6.

[17] *Kenkon Bensetsu*, pp. 10, 16, 18, 20, 23, 29, 39, 45, 48, 54, 63, and 97.

[18] *Kenkon Bensetsu*, pp. 16, 29, 39, and 61.

[19] *Kenkon Bensetsu*, pp. 38, 58, and 85.

Mukai was the first who explicitly admits to Western superiority at the physical level, and his observation discussed above is the first instance of Japanese recognition: the Westerners are good at a physical level, but ignorant of discussions at a metaphysical level. We know that Arai Hakuseki (1657-1725), a Confucian scholar and advisor to the sixth shogun, had a similar impression on the investigation of the captured missionary Giovanni Battista Sidotti (1668-1714). And, of course, the idea culminated in Sakuma Shozan's (1811-64) famous dictum: "Eastern Morality, Western Science and Technology". This idea facilitated the Japanese in justifying the introduction of Western science and technology which are considered to be superior, while they could maintain their identification at a metaphysical level at the same time.

6. Other Manuscripts

It is extremely difficult for us to estimate how widely the *Kenkon Bensetsu* had been circulated, because we find only three sets of its copies extant.[20] A few words should be mentioned about the *Nigi Ryakusetsu*[21] (A Brief Discussion on the Two Worlds, [Celestial and Terrestrial]) which deals with the same subjects. The author Kobayashi Kentei (1601-83) is said to have been imprisoned for twenty-one years because his teacher was suspected of being a Christian. He was set free in 1667 when he was sixty-seven years old. Therefore, the book under discussion must have been written between 1667 and 1683, although the only extant manuscript says that copying was completed in the autumn of 1715. The *Nigi Ryakusetsu* follows more closely the texts of Gomez's *de sphaera* than the *Kenkon Bensetsu*. The order of chapters is more or less the same: the parts on the celestial world come first, and then the parts on the sublunary world follow, which is a quite opposite order in the case of the *Kenkon Bensetsu*.

---

[20] Before World War II, three more sets of copies existed.

[21] The manuscript is preserved in the Cabinet Library (Naikaku Bunko) of the National Archive, Tokyo. The text is available in a recent print: *Nihon Shiso Taikei*, vol. 63 (Kinsei Kagaku Shiso, ge [Scientific Thought in Edo Japan, the second part] (Iwanami Shoten, 1971), pp. 10-107. See also Obara Satoru, "Yoroppa kagaku shiso no denrai to juyo [The arrival and acceptance of European scientific thought]", *ibid.*, pp. 481-496. For a brief discussion about the book, see Nakayama, *A History of Japanese Astronomy*, pp. 98-100. The man who copied this work named Ohe notes that the manuscript he referred is the autograph of Kobayashi. But this does not seem to be the case, for the order of chapters are somewhat confused. For example, the chapters for four elements are inserted in the part for the celestial world in this extant copy, although the editors correct this in a recent print.

Probably Kobayashi based his renderings on the Japanese translation of Gomez's treatise on the sphere.

There remain similar extant manuscript texts which seem to be different versions of the *Kenkon Bensetsu.* This suggests that different versions were made to meet the needs of the readers, but hand-written copies were perhaps intentionally destroyed, fearing that the mere possession of the work written by an ex-Christian father might bring about severe punishment.

There are four titles which have their own characteristics.

(1) *Nanban Unkiron* (A dicussion by the Southern Barbarians on the *yun qi*)[22] (preserved by the Ohkochi family; its new hand-written copy is available in the Japan Academy).

According to the note written by a certain man Sugi Teian in 1670, this was copied from the manuscript possessed by Matsudaira Terutsuna, who went to fight in the battle against the Christian uprising in Shimabara (1637-8). This experience led him to take an interest in the art of war and eventually to write a book on firearms and how to enter into battle for foot soldiers. From the same interest, he seemed to have acquired the manuscript on astronomy transmitted by the southern barbarians. Two different points can be detected in comparison with the *Kenkon Bensetsu.*

(a) In the *Nanban Unkiron*, the astronomical part comes first and the explanation on terrestrial phenomena as well as four elements follow, the order being quite opposite to the *Kenkon Bensetsu*, and the same sequence as in the *Nigi Ryakusetsu* and Gomez's original text.

(b) Each passage is shorter than the one in the *Kenkon Bensetsu.* We can also find quotations from Chinese classics which cannot be found in the *Kenkon Bensetsu.* Quotations are from *Nei Jing*, a classic medical publication; the *Book of Change* and Confucian *Annalects*. Together with the term *yun qi* in the title, and the fact that *Su Wen* in the *Nei Jing* is famous for its insertion of the theory of *yun qi* in the Tang dynasty, the existence of a chapter entitled "On blood, phlegm, yellow bile, etc." suggests that this piece of work may have been meant for medical practioners. It is quite interesting when we remember that Chuan (Ferreira) worked as a medical practitioner and was sometimes summoned to the Dutch factory on Deshima for consultation.

---

[22] Cf. Ohya Shin'ichi, "*Kenkon Bensetsu* no ichi ihon — *Nanban Unkiron-*" (Another version of the *Kenkon Bensetsu — Nanban Unkiron-*), *Kagakushi Kenkyu,* no. 14 (1950), pp. 35-39.

(2) *Nanban Tenchiron* (A discussion by the southern barbarians on the heavens and the earth) (Preserved in the Kano Collection, the Library of Tohoku University).

The text is the same as the first: *Nanban Unkiron*. One point which is worth mentioning is the seal at the very end, which reads "Soko". If "Soko" in the seal refers to Yamaga Soko, a famous Confucian scholar, it is quite interesting, for it has never been claimed that Yamaga Soko took any interest in astronomy.

(3) *Tenmon Biyo* (Compendium for Astronomy).

The following differences are to be pointed out, compared to the Kenkon Bensetsu:

(a) this hardly discusses the four elements;

(b) discussions on meteorological phenomena, tides and earthquakes are missing;

(c) there is no reference to climates.

We can conclude from the above points that this version concentrates only on the heavenly sphere in Aristotelian dichotomy.

(4) *Shidai Zensho* (A Complete Work on Four Elements) (preserved in Kyushu University).

I have had no chance to consult with this manuscript. We should note, however, that none of these manuscripts have Mukai's commentaries.

PORTUGAL AND KOREA: OBSCURE CONNECTIONS IN THE PRE-MODERN SCIENCES BEFORE 1900

PARK Seong-Rae
Hankuk University of Foreign Studies, Seoul

I

In the beginning Portugal was 'Bullanggi' in Korea, when the Koreans had known, for the first time, anything about the country in 1520. 'Bullanggi' is, of course, the Korean rendition of the Chinese word for Portugal: Fo-lang-chi (佛朗機), the term we can also find from the History of Ming (明史). The first incident from the history of Korea is the one in 1520, when Korean interpreter Yi Sok (李碩) had returned from his official visit to China, and delivered his report to the throne, where he mentions about Bullanggi. His report goes something like this:

> *"Ever since the opening of the Ming Dynasty, the Bullanggi* (佛朗機) *could not pay tribute to China because the route of transportation was blocked by the Manra State* (滿剌國, Malacca). *Now that the Bullanggis had destroyed the Manras, they came to China to request the enfeofment from China. But after some discussions the Chinese Ministry of Rites answered that China could not grant it, because they had wilfully destroyed the state under Chinese recognized enfeofment. Though they were refused the formal imperial audience, their waiting at the pavilion in Beijing was all the same as the other embassies visiting China."*

According to his report, they looked somewhat like the Japanese, but their costumes and culinary customs were so unhumane that the Chinese would say they had never seen such a (barbarous) people. But when the emperor would go out of the palace, he would bring a group of foreigners composed of two or three men each from the embassies of the Tartars, Muslims, Bullanggis, Thais, and Lhamas. Sometimes the emperor had even tried to learn some words from their languages as well as watching their performances.

Obviously this Korean interpreter during his visit to China must have actually seen the Portuguese people in Beijing. It was the first time that any Korean in Korean history had seen a Westerner and had recorded it. Of course that was not in the Korean soil, but in China, and the meeting was not much of a meeting in a sense. In that year Korean government had sent its embassy to China as a usual ritual of ceremonial visit for their part. The embassy in that year was headed by Sin Sang (申金常) and Han Hyowon (韓效元) most probably made of some hundreds of the members including the interpreter introduced above. At their return to Korea, they were met by the King, who had asked about any of the new countries visiting China in that year.

To this question they answered about the Bullanggi (Portuguese) in China. Their country occupies the southwest of the Western Region of the world, according to the report. Their embassy also stayed (with the Koreans) in the Yu-he-kuan (玉河館). And the embassy was made of twenty some members under one ambassador. After having contacts with them, Korean visitors found them as having open and civilized minds. Their books show their characters, which were very similar to the Sanscrits or the Korean alphabets, but very thin and accurate. Their clothes, made of materials weaved from the geese feathers, were round-shaped with wide bottoms, and they put on them from their heads without any parts to tie or untie. For their eating, they have only chicken and noodles, because they are the only products in their land. When the Koreans asked about their customs, they answered that even their ruler have only one wife, and when the wife dies he does not marry again.

When the king of Korea asked about the Chinese response to them, their answer continued:

> *"When they had arrived in China, they had complained much about the dirtiness of the Yu-he-kuan. The Chinese Ministry of Rites did not like their rudeness, and even after three years they were not well treated. They had brought plenty of gold and silver, so much so that they would buy anything with gold and silver. When the Koreans had visited their place, the place was fully covered with colored tents. Inside the place chairs were arranged in both sides*

of the east and the west. And in the middle of them was placed one chair covered with a red blanket. According to their explanation, this was the one for the emperor, when he would pay a visit there. When the emperor would encounter with the tributary mission, he would visit the pavilion, according to their explanation. The Chinese side had also admitted that the emperor would surely visit their place when he would return to the capital from his sojourn. When the Korean king asked about their route of travel to China, the ambassador answered that they came to China all the way through the seas, and after their landing at Kuang-jou they have traveled 3 thousand some li (里: Korean mile) *to the capital."*

II

After the record, we do not have any information about Korea's direct relations with Portugal in Korean history for hundreds of years thereafter. I could not find any sign of the Portuguese visits to the Korean soil in the traditional period, nor could I read out any Korean contact with the Portuguese. But the familiar expression "Bullanggi" appears repeatedly in the Korean historical records in the period. Somehow the same expression of Bullanggi, however, does not mean the country of Portugal any more. Instead the expression is solely used to indicate the western style gun thereafter. Ever since the Ming army in its final years had decided to adopt the Western guns brought from the Portuguese, or more aptly from the Jesuits in Macao during the last days of the battles against the invading Manchus in the early 17th century, the word, originally meant for the country of Portugal (Bullanggi or Fo-lang-chi), became settled as their guns, or their styled guns in Korea.

Yi Su-gwang (1563-1629) had his answer for this. According to him, the same expression Bullanggi was the name of a country on the one hand, but it often used to mean the gun made by the people of the Bullanggi. In other words, the Bullanggi was Portugal for the Koreans in the seventeenth century, as much as the Portuguese guns at the same time. Since the Koreans did not have any more actual relations with the Portuguese thereafter, the word had remained to be used only to indicate the guns, the originally western style guns produced actually in Korea thereafter.

Koreans had tried to make the guns for themselves from the early period, which I could not at this moment decide exactly when. The earliest record in the *Veritable Records* (Sillok, 實錄) of the Choson Dynasty tells us that Korean officials proposed to forge the bullanggi — "the most effective" guns — from 1681 on. Not only the most effective, but they were probably considered as the biggest kind of artilleries for the traditional Korean army in the earlier times. And eventually the bullanggi were classified in the later period into five different sizes.

Today there are a lot of them in differing sizes preserved and displayed throughout the museums in Korea.

III

In his book, *Chibong Yusol* (芝峰類說, Classified Writings of Chibong), Yi Sugwang has a list of 35 "foreign countries" of his days, which include only a few definitely European countries such as Portugal, England, and Europe (as a separate country just like England or Portugal). Yi Sugwang had this comment about Portugal. According to him, Portugal is a big country located in the southwestern seas of Thailand. The guns of the country were known as the bullanggi, which had been used by the Korean army. Also produced are the so-called Western textile, that is very light and as delicate as the wings of cicada. Almost the same comment about Portugal is found from another scholar's writing later, in An Chongbok's (1712-1791) *Chaptongsani*.

But I could not find any Portuguese arriving at the shore of Korean peninsula during the long period of three hundred some years of the Portuguese contacts with the Chinese and the Japanese. In other words no single Portuguese had arrived at Korea until the nineteenth century. At present it is believed that a Dutch named Weltevree was the first westerner who had come to Korea in the traditional period. He had arrived to Korea in 1622 to naturalize himself as a Korean with the Korean style name of Pak Yon (朴燕), and eventually to serve for the Korean army as an artillery specialist. He married a Korean girl and had children. But historians had not yet been able to find any information thereafter about this first Westerner in Korea.

Indeed he was the first of the many Dutchmen who had been shipwrecked at the Korean shores ever since. The most famous event of the flying Dutchmen in Korean history is Hendrik Hamel's shipwreck in Korean seas and his escape from Korea in the mid-seventeenth century. Hamel's crew of 38 members, most of them must have been Dutch, had arrived at the shore of Cheju Island off in the southern sea of Korea in 1653. Obviously they could not identify them as any of the known countries. But Weltevree had met them to identify them as the "southern barbarians", thus indicating the people of his fatherland.

Hamel's adventure is very well documented today, thanks to his memoirs published later after his escape in 1666 from the Land of Morning Calm. I could come up with a rare document about the first Western words transliterated into Korean alphabets from a Korean book written in the early 19th century. In his collection of miscellaneous writings, *Chuyong-pyon*, Chong Tong-yu (鄭東愈, 1744-1808) gives us a precious record about a possible Portuguese relations in 1801, when five Westerners had arrived in Cheju Island. According to this narration, they are from Macao with the names of Bunansigo (22), Yollinandu (25), Andrusu (24), Mariandu (32), and Ggoidanu (33), the last two men being Negroes. There are 101 Western words transliterately copied down in Korean alphabets. Though it is still not too clear whether they are Portuguese words, or Spanish, they did not give any influence whatsoever to the Koreans at the time.

According to the record they were sent to China immediately after they had arrived at the Korean shores. But the Chinese would not accept them to return to their land, because the Chinese are not like in the olden days, when they were very kind to the foreigners. They simply answered that they would not know the place of their origin, and the Korean embassy had no other way but to bring them back to Korea, and to send them to the original place of their arrival. According to the recent news to the author, two of them had already died, leaving three behind when he was writing this book in 1805-6.

This story is documented also in the Sillok several times. According to the local governor's report to the throne, one of the five died on his way to China in October 1801, the other passed away in 1805 for illness. And the local governor had discovered from the ship-wrecked Ryuku people that they are from Yosong (呂宋, Luzon). So the Koreans earnestly requested the other ship-wrecked Ryukyuans to bring them back to their country on their way home. But the Ryukyuans did not comply to the demand. So the governor suggested to the throne to send them to China once more, since their country of origin was identified this time. Korean king decided to do that only when the annual ambassador from Korea receives an affirmative answer from the Chinese court on his next visit to Beijing. But I could not find out any more information about the matter thereafter.

Several more records we can find about the Dutch visits, but none of the Portuguese or Macao connections after 1809.

IV

Before the Korean connections with the Dutch began in the mid-seventeenth century, Koreans might have had some other contacts, though indirect, with the Portuguese. According to the documents of the Catholic church for the period, the Portuguese priests in China and in Japan had attempted to make some kind of meaningful contacts with the Koreans in the Korean soil by sending some of the Portuguese fathers there.

The first case of the Portuguese interest in Korea is found from the Portuguese missionaries who had stationed themselves successfully in the Western part of Japan in the mid-16th century. Jesuit father Luis Frois was the first foreigner, who had written, though piecemeal, about Korea in his enormous volume of the history of Japan (*Historia de Japam*). Though I do not have any concrete evidence to support it, Father Frois might have gathered his information about Korea directly from the Spanish missionary, Father Gregorio de Cespedes, who had been in the Korean soil during the Hideyoshi Invasions to Korea in the years 1592-98. Cespedes had been rather well known among Korean historians as a "Portuguese" missionary, until one of my colleagues discovered his place of birth as Madrid instead. Whether he was a Portuguese or a Spaniard, Cespedes must be the first known Westerner who had actually visited Korea for the first time. He stayed in Korea for a period of slightly more than one year under the military camp of a Japanese Christian general, Konish Yukinaga (小西行長), during 1593-4. He is believed that he had successfully Christianized hundreds of Koreans in the mean time, and many of them followed him back into Japan during and after the war.

The war was devastating for all the three countries of East Asia at the time. The aftermath was really tragic particularly for those Korean prisoners brought to Japan toward the end of the war. According to the contemporary records of the Catholic church, the Portuguese priests in Japan at the time had tried to save these wretched Koreans, and had actually succeeded in doing so for several hundreds of them. And at least once in

August 1, 1598, they had a meeting among themselves to discuss the matters in Nagasaki.

From China, on the other hand, none other than Hsu Kuang-chi (徐光啓), the most illustrious colleague of Matteo Ricci, had once volunteered to go to Korea for the evangelistic purpose. Actually he asked in 1619 to the throne to dispatch him as the Chinese ambassador to Korea. His plan for the propagation of Christianity in Korea included the accampaniment of the Jesuit Italian priest Francisco Sambiaso (畢方濟), and a tactical plan to stop probable invasions of the Japanese from the seas by building a strong fortress in the southern coast of Korean peninsula under the control of the strong guns of the Portuguese. The plan was ultimately ended as an abortive dream, because the other party of the Jesuits recommended the Chinese emperor not to send such an important person to Korea in such a dangerous time of impending foreign threats.

The first baptism to a Korean was also given by a Franciscan priest Alessandro de Gouvea (湯士選) of Portugal in Beijing in 1791. Bishop Gouvea had also left a valuable document about the status report about "the introduction and progress of Christianity in the peninsula of Corea from 1784 to 1794". And the first Korean baptized by him was Yi Sunghun (李承薰, 1756-1801), who had returned to Korea with several books on science — including the Elements (幾何原本), another mathematical book (數理精蘊), a telescope, and a western style sun-dial (地平表), as well as T'ienchu shii (天主實義).

And finally the Korean Catholics had very meaningful connection with the city of Macao, for they have their first priests trained in Macao from 1837. Originally the Western priests had smuggled themselves into Korean land in 1835, for the first time after Cespedes in the 1590's. One of the priests had immediately returned to Macao with three Korean boys recruited to be trained as Catholic priests. They were Kim Taegon (金大建, Andreas, 1821-1846), Choe Yangop (崔良業, Thomas, 1821-1861), and Choe Pangje (崔方濟, Francis Xavier). The last of the three Koreans had died from fever in 1838, while two of them had successfully finished their training as Catholic priests for the first time in Korean history. I do not know yet about whether their training in Macao had any relations with Portuguese priests at the time in Macao, or how they might have been educated in the western sciences.

At any rate the first Korean Catholic priest, Father Andreas Kim, naturally revered by all the Catholics of Korea today, has been remembered now by the people and visitors of Macao, as his statue stands in the Camoes

Garden in the downtown of Macao. Only one year after his permanent return from Macao in 1845, Father Kim at the age of 26 was arrested and beheaded on July 26, 1846.

V

Portugal seems to have little direct relations with Korea in terms of Korea's absorption of Western science before 1900, although the Christian connection is very much notable and duly recognized among the Koreans. But Italian Jesuit Ricci's influence in Korea is very well known among the Korean historians. Of course, few Koreans would have knowledge that Ricci, though himself an Italian, had actually departed Lisbon, Portugal, in 1578, to arrive at Macao in 1582.

But the most important single incident in the Korean history of science in terms of Korean connection with the Portuguese is the Koreans' contact with a Portuguese priest named Joao Rodrigues (陸若漢, 1561-1634) in the early 1630's. Today every school child learns about the first introduction of the Western things into Korea in 1631. This is the most celebrated Western contact of the Koreans in Korean history, and that is Ambassador Chong Tuwon's (鄭斗源) importation of the Western things from China, including a telescope and a Western style gun, as well as several kinds of books and a world map. The books include *T'ien-wen-lueh* (天文略) by Emmanuel Diaz (陽瑪諾, 1574-1659), a famous Portuguese missionary of the time.

More important than this importation of Western things, Rodrigues had actually had some intellectual contacts with a knowledgeable Korean named Yi Yonghu (李榮後) in Teng-jou (登州), Shan-tung (山東), China, at that occasion. Among the very scant record of the period about the Korean contacts with the West, we are very fortunate to have the detailed records of the correspondence between these two men of knowledge across the two countries which had never had any intellectual contacts until that time. Already trained with some Western astronomical knowledge in Korea, through the books written by the Western missionaries in China, Yi Yonghu asked several matters to Rodrigues by writing.

His questioning to the Portuguese priest includes whether China, despite of the global shape of the earth, could be considered as the "central" kingdom, as the country's name would have it. To this question of the Korean, Rodrigues answered that every country could be seen as the center of the world because there were many countries on the surface of the globe, which is the earth. His answer, though trivial for us today, was of cardinal significance to the minds of the Choson kingdom at the time, when almost every Korean had thought that the "middle kingdom" (中國) was indeed in the geographical center of the world.

The Korean left his record about his impressions about the Western astronomy at the times, which include the theory of the 12-layers of the heavenly spheres, as well as the exactness of the Western calendrical science. He showed his amazement about the many countries and peoples on the distant parts of the globe, which the Koreans did not have even imagined until that time.

Rodrigues is a Portuguese priest who was active in Japan in the earlier period of his life. At the age of 16, he came to Japan in 1577, and very soon he must have mastered Japanese to become a teacher at the Shimabara Seminario. With his command of Japanese he had at least once served as the interpreter between a Jesuit priest and Toyotomi Hideyoshi. Though his good relations might have been preserved even with the Tokugawa government thereafter, he was not free from the early Tokugawa persecutions of the Christians, and he had finally left Japan in 1610 to seek an asylum in Macao. Before he left Japan, however, he was already the writer of *Dicionario da Lingoa Japonesa* and *Historia da Igreja do Japano*.

And it was while he was in Macao that he was recruited by the Chinese military commander Sun Yuan-hua (孫元化) in Shan-tung, who happened to be a disciple of Hsu Kuang-chih and himself a Christian. And in that capacity Rodrigues had met with Korean ambassador Chong Tu-won and his assistant Yi Yonghu in 1631. It is obvious that he was very good with Japanese, but we can find no evidence for his versatility about the Western science.

The Koreans, for the first time in its history, at the time of 1631, were put to the mercy of this novice scientist from Portugal. And this is about all we can find out about the Portuguese-Korean connections in the history of sciences before the nineteenth century. And it was also the only direct Western connection in terms of science on the intellectual level. It is surprising indeed that Korean history shows this much poverty in the intellectual contacts with the West in the traditional period. The contrasts of the Korean experiences with those of the Chinese and the Japanese in terms of their relations with the Portuguese are really impressive.

REFERENCES

Chungjong sillok (中宗實錄) 16: 6 (15年12月14日 戊戌).

Yi Sugwang. Chibong yusol (芝峰類說) 卷 2, "外國".

Ibid.

An Chongbok (安鼎福). Chaptongsani (雜同散異) 卷 4, 123-129 "芝峰類說 諸國部 抄".

Yi Byongdo (李丙燾)(譯註) Hamel p'yoryugi (하멜 漂流記) Seoul: Inchogak, 1954.

Ibid. Also Ledyard, Gari. The Dutch come to Korea. Seoul: Royal Asiatic Society-Korea Branch, 1971. ハメル, H. (生田滋 譯) 『朝鮮幽囚記』 東京: 平凡社, (東洋文庫).

Chong Tongyu (鄭東愈). Chuyong-p'yon (晝泳編).

Sunjo sillok (純祖實錄) 7(1807)/8/10 & 9(1809)/6/26.

About Frois see Yi Kyut'ae, Choson kyonmun, "Porutugal saje-ui Imran kyonmun-gi" pp. 9-41; and Luis Frois, Historia de Japam (in Portugese); or Frois, P. Luis, s.j. Die Geschichte Japans: 1549-1578, Leipzig: Im Verlag der Asia Major, 1926. Nach der Handschrift der Ajudabibliothek in Lissabon; uebersettzt und kommentiert von G. Schurhammer und E.A. Voretzsch. "Einleitung"; also Japanese translation in フロイス(松田毅一, 川崎桃太 譯) 『日本史』 全12巻 東京: 中央公論社, 1980.

Pak Ch'ol, Ssesupedesu (Cespedes). Seoul: Sogangdae Press, 1987.

Dallet, Histoire de l'Eglise de Coree, Paris, 1874.

Kang Chae-on. Soyang kwa Choson. Seoul: Hakkojae, 1998.

Chongjo sillok (正祖實錄), 33:58b (15 11), Hong Isop, Choson kwahaksa, Seoul: Chongumsa, 1946, p. 261.

TRANSLATIONS OF PORTUGUESE TEXTS INTO KONKANI AND KONKANI COMPOSITIONS INTO PORTUGUESE WITH EDUCATIONAL INFLUENCE ON LITERATURE AND ART AND THE TRANSFER OF TECHNOLOGY

Joseph VELINKAR
Heras Institute, St. Xavier's College, Bombay

The coming of the Portuguese to India five centuries ago, a major event in world history, led not only to the discovery of the sea-route to the East with the landing of strange men and the anchoring of their ships on distant shores, but the safe return of the crew, Vasco da Gama in particular, and the exchange of desired goods, merchandise and knowledge, the last of which is the object of this research paper.

Many were the channels of exchange between two very different peoples, one of the most important being the means of communication. On reaching Calicut in 1498, the Portuguese met two Moors from Tunis, who spoke Spanish and Turkish.[1] These became their first interpreters. The Portuguese soon discovered that one of them spoke their language fairly well.[2] Years passed and Afonso de Albuquerque conquered Goa. Many Portuguese began to settle in Goa, which even merited a bishop. When Francis Xavier came to India as Papal Legate in 1542, he began to teach catechism to the Portuguese, other Christians and their slaves in a language which can be called "Hispano-Portuguese". For those who did not understand this language, he chose an interpreter from among the bystanding

---

[1] Herculano, A., e de Paiva,o Barão do Castello (eds.), *Roteiro da Viagem de Vasco da Gama,* 2ª edição (Lisboa: Imprensa Nacional, 1861), p.51. (=*RVVG*) .
[2] *Ibid.*, pp.51-52.

listeners, but specially from among the children.[3] By the year 1545, the grammar students of Holy Faith Seminary were preaching in their native Konkani. Two of them preached every Sunday in two churches.[4]

This led to the next stage — the foreign priests (or to-be-priests) themselves wishing to study Konkani. Accordingly, a Konkani Language School was set up at St. Paul's College (the new name of Holy Faith Seminary), Old Goa. It was then the teachers felt the acute need of a grammar (*arte*) and dictionary (*vocabulario*). They set about their task of composing these immediately, finished the work by 1569 and began using them in Old Goa. These can be called the first Portuguese–Konkani works used in Goa. Many other Konkani grammars were composed over the years and the next century. Some of the better known are:

1) *Sintaxis Copiosissima* (c. 1635) by Fr. Gaspar de S. Miguel, a Franciscan,
2) *Arte Grammatica da Lingua Canarina* (c. 1635) by Fr. Christovão de Jesus and
3) *Arte da Lingoa Canarim* (1640) by Fr. Thomas Stephens.

The Jesuits also composed a dictionary, improved upon by several later authors, entitled *Vocabulario Da Lingoa Canarina Com Versam Portuguesa* and published by the Junta de Investigações (Lisbon) in 1973. For these and some of the religious books composed during this period, I would refer the reader to *The Ancient Franciscan Provinces in India* by Achilles Meersman, published in Bangalore in 1971 and a recent book *India and the West: The First Encounters* by Joseph Velinkar published just this year by the Heras Institute.

Many works were original compositions in Portuguese on religious topics or translations from the Latin. To mention only a few: Fr. Antonio da Costa, a Jesuit, composed an unpublished work "Tratado de como se hão de catequizar os novamente convertidos". This work was lost in time. Another work by an unknown Jesuit has been entitled "Catecismo ou Práticas da Doutrina Cristã" on the articles of Faith, Commandments and Sacraments.

---

[3] Schurhammer, Georg, *Francis Xavier* II, (tr. by M. J. Costelloe S.J.), (Rome: The Jesuit Historical Institute, 1977), pp. 222-23.

[4] Wicki, J. (ed.), *Documenta Indica* I, (Rome: Institutum Historicum Societatis Jesu, 1948), p.46. (=*DI*).

This work too was never printed.[5] Another unpublished catechism in Konkani was written by the Franciscan Manoel do Lado.[6] Several of them, like the famous Catechism of Francis Xavier composed in Portuguese and printed posthumously as *Doutrina Cristã* in 1557, were original compositions. Another famous Portuguese catechism of this period was composed by Fr. Marcos Jorge, printed in 1566 in Portugal and sent to India. It was translated into Konkani by the famous English Jesuit Thomas Stephens sometime after 1585. By 1614 the Konkani version of the Jorge catechism appeared in print and stole a march over the Portuguese original. Very soon this translation was used in all the parochial schools of Salcete.[7]

Between the Jesuits and Franciscans, several grammars in Konkani (and even in Malayalam and Tamil), Portuguese–Konkani dictionaries, catechisms and pious books — some in Portuguese, some in Konkani and some translated into Konkani from Portuguese originals — were composed. Among the manuscripts translated into Portuguese directly from the Marathi were two troves recovered from the house of a Hindu gentleman in 1558. The works were translated into Portuguese by a certain Manuel and sent to Rome. They now lie in the Jesuit archives under Codex Goa 46, 348-94 and some of them as second copies in the Biblioteca Pública at Évora, Portugal. The translations were scheduled to appear in the *Archivo Italiano per la Storia della Pietá,* Rome.[8]

In the 18th century, John Ernest Hanxleden, a German Jesuit, wrote among his numerous works a Sanskrit–Portuguese and a Portuguese–Sanskrit dictionary. A copy of these works lie in the Vatican Library.[9]

Towards the end of the 17th century all this came to an abrupt end in Goa. The main reason for this decline and stoppage of exchange between the Indian languages and Portuguese was the decree passed in 1684 by Viceroy Francisco Távora banning the use of Konkani within three years and prohibiting the celebration of marriage and sacerdotal ordination of those who did not know and speak Portuguese, thus making Portuguese compulsory for all.

---

5 Saldanha, Mariano, *Doutrina Cristã em lingua Concani por Tomás Estevão,* (Lisboa: Agência Geral das Colónias, 1955), pp. 1-10.

6 Meersman, Achilles, *The Ancient Franciscan Provinces in India* (Bangalore: Christian Literature Society Press, 1971), p. 85.

7 *DI* VII, p.385; Gonçalves, Sebastiam, *Primeira Parte da História dos Religiosos da Companhia de Jesus* III (Coimbra: Atlântida, 1962), p.100. (=*PPHRCJ*).

8 *DI* XII, pp. 23-26.

9 Amaldass, Anand, *Jesuit Presence in Indian History* (Anand, India: Gujarat Sahitya Prakash, 1988), pp. 182-204.

Three years later the *alvará* dated 17 March 1687 was signed, confirming legislation and signalling the end of the golden years when Konkani and Portuguese lived in harmony.

The big contribution of Portugal to India in the field of technology is the printing-press. The first printing-press machine in India came by a strange convergence of circumstances. In 1556 the Patriarch of Abyssinia João Nunes Barreto arrived with it at Old Goa *en route* to Ethiopia. The unassembled machine was of course a novel spectacle. As it was not possible at the time to proceed to Ethiopia, the Patriarch remained in Goa, and so did the machine. A Spanish Jesuit, Bro. Juan de Bustamante, and his Indian assistant, who had come with the Patriarch, set up the press and began to operate it. This press brought out several works, including the *Doutrina Cristã* of Francis Xavier in 1557, which is the first booklet printed in India. All the known works printed in Goa were in connection with religion or religious needs. After this first press set up at St. Paul's College, another belonging to John Quinquencio and John Endem came up. But hardly anything is known about it. Goa's third press was set up by the Jesuits at St. Ignatius College, Rachol, where the famous *Purana Cristão* of Fr. Thomas Stephens was printed. — This work written in Marathi/Konkani is considered today a classic Marathi literary work and old copies of it have been preserved by Christians on the west coast of India. — With the press, there came out to Goa expert printers, like Bro. João Gonsalves who worked "with iron and metal, making moulds of letters for printing and other delicate things". Before his death in 1579, he is thought to have printed books in the Tamil and Devanagari scripts.[10] Fr. João de Faria (†1581), "prefect of the printing-press and of the works" was known to have tried printing with Devanagari or Halakannada characters".[11] Besides books, holy pictures were also printed. The one known example is a picture of the crucified Christ with his mother and the apostle John on either side printed at the request of Patriarch Barreto for distribution in 1560. At first only a few dozen, but in the following month some 8,000-10,000 came out of the press.[12] But the reams of paper and pitch had to be obtained from Portugal.[13] It is difficult therefore to say how far there was a transfer of technology and Indian printers would have been able to manage on their own.

[10] *DI* XI, p. 652.
[11] *DI* X, p.1006, fn. 53.
[12] *DI* IV, p.798.
[13] *DI* XI, p.588.

The existence and running of a College made another technological demand on the Jesuit management. Already in 1549 Fr. Antonio Gomes, Rector of St. Paul's College, had written to King João III expressing the need of timepieces for the better functioning of his institution and a Jesuit house in Bassein (today called Vasai). By 1552 they had received a timepiece and the College routine was regulated by it.[14] In 1567 the timepiece was corrected by calculating the time at which a lunar eclipse would begin.[15] Not long after, timepieces began to be found in Goa.[16] Jesuit mechanics were able to repair and make timepieces and clocks.[17] One of these watch repairers was Fr. Thomas Stephens, Rector of Holy Spirit College at Margao, Goa.[18] In 1604 the reconstructed Margao church was enlarged. It had two houses attached and a clock tower.[19] In 1734 it was noted that a Jesuit even had a gold watch entrusted to him, probably in Portugal.[20] Thus it has been established that the Portuguese introduced watches and clocks into Goa, but how widespread their usage was has not yet been determined.

The next area of influence was architecture. The Manueline style of architecture and the later Baroque style with hall, transept, apse and façade with one or two bell-towers planned by Portuguese architects and executed by Indian masons, craftsmen and artists in brick and stone used in Goa during the 16th century and later are visible there even today. Strikingly singular are the multi-storied structures with a gallery running round the length and/or breadth of the church. The two-storied façade of St. Francis Assisi church, the three-storied façade of the Bom Jesus church (and residence) decorated with gilt intersecting vaults and glass windows, and the five-storied tower of the Augustinians — all in Old Goa — are only some examples. All this construction required technical skill and expertise. This new style is seen, in greater or lesser degree, in all the churches of Goa. This they learned from the Portuguese architects and engineers: Ambrosio Argeiro, Julio Simão, João de Faria *et al.* They trained others, like Bro. Francisco Aranha and the *mestre*

---

[14] *DI* I, p.528; II, p.342, fn. 12.

[15] *DI* VII, p.176.

[16] *DI* X, p.427.

[17] *DI* XI, p. 652.

[18] Rome, Archivum Romanum Societatis Jesu, Goa XV, f.174b, ll. 7-9.

[19] *PPHRCJ* III, p.91.

[20] Lisbon, Torre do Tombo, Jesuitas, Armario 81, f.10.

at Chaul,[21] who were then able to plan and construct on their own. The new style and technology are seen in the construction of forts, for example in Mormugao and Rachol with its ramparts, moats and bastions, all remnants of the European castle, and in several houses that still stand in Goa, Chaul, Damão and other places.

Two of the greatest contributions in the biological field are the planting and grafting techniques. The mango seems to be a fruit and tree long grown on and inured to Indian soil. But the Portuguese came and, with the grafting technique, created some of the top varieties of mango, all bearing Portuguese names. The best mango variety, most in demand today, is the *affonso*, popularly known in Indian languages as *hapus*. It is the most expensive variety and is being exported to several countries. Other well known varieties are: furtado, monserrate, fernandina, malcorado and a list of others. In the Central Library, Panjim (Goa), a book titled *Arte Palmarica* explains the technique of growing more and better coco trees in a given space and the manner of growing them. The anonymous work is attributed to a Jesuit Brother who cultivated the farms entrusted to his care. The work and the knowledge implied seems to have made a big difference to the Portuguese coastal territories. An anonymous traveller voyaging along the west coast of India remarked that the stretches of land which had come under Portuguese rule at some period of time or other are marked with denser coco-tree-cultivation than other territories along the same coast.

Over and above this, the Portuguese imported many flowers, fruits and trees to Goa, transforming the territory into a veritable garden. The exotic fruits and flowers are numerous. To mention only a few outstanding ones: the cashew, pineapple, custard-apple and pomegranate together with the shoeflower, sunflower and tobacco were all brought by and under the Portuguese from America, the West Indies, Africa, West Asia and the East.[22]

Another very important change of art and technique was in shipbuilding. In the early years of Portuguese rule and expansion in Asia, when Vasco da Gama landed at Calicut and Afonso de Albuquerque conquered Goa and was faced with a counterattack, the Portuguese could always retreat to their ships and sail away in safety. But they could not always depend on home-built ships, which sometimes reached the East in a battered state.

---

[21] Barros, João de, *Da Asia*, dec. VIII (Lisboa: Livraria São Carlos, 1973), pp. 420-21. (=*DA*).

[22] Goa, *O Oriente Português* (Journal of the Archaeological Commission of Portuguese India. Old Series I-IX ,– Nova Goa, 1904-20; New Series: Goa: Tipografia Rangel: 1931-41) VI, pp. 224-28.

They found good teak in Bassein and soon took to shipbuilding in India. The shipyard set up in Old Goa was capable of producing strong and beautiful ships, the most famous of which was the *nau* 'Cinco Chagas' constructed in Goa in 1559-60 at the personal expense of Viceroy Constantino Bragança. She served the *Carreira* 25 years, making 9-10 round trips, and was the flagship of five Viceroys.[23] It was felt that ships built in India were certainly stronger, but not always cheaper. The main shipbuilding yards in India were at Goa, Bassein, Cochin and to some extent in Daman.[24] It is not surprising therefore that in the next centuries the Marathas should be asking the Portuguese to help build their ships in order to mount and withstand attacks on and from the enemy.

I will now conclude this first part, not with clear transfers of technology but only with the mention of improvements in many other fields of life: armaments, guns (made in Cochin, Quilon and Sri Lanka),[25] horses, quality of wood, style of painting, music, musical instruments, domestic architecture and a host of other items which make for the improvement of the quality and comfort of daily living, though it must be stated that many of these changes did not spread easily both in and out of Portuguese India

As far as the transfer of technology from India to Portugal and the West is concerned, there seems to be little to show, the chief reason being that writing was done on a limited surface and scale. There were several stages of learning to write: the last being tracing letters on a palm leaf with a stylus. The palm leaves were not preserved usually. What had to be preserved for a long time was carved on a copper plate or on stone. A number of palm leaves have been preserved, but on these surely no maps could be drawn. Whatever knowledge was learnt, was passed down the generations by word of mouth or by demonstration in action, as Vasco da Gama learnt when he brought a Christian pilot aboard his ship.

However, we do find references to medicines or drugs, even a shipload of them, being sent to Portugal.[26] The knowledge of medicine was communicated only by word of mouth in Goa. Garcia de Orta came to India and put down in his *Coloquios dos Simples e Drogas da India* the information gathered in the country. Similarly several Portuguese navigators coming to

---

[23] Mathew, K. M., *History of the Portuguese Navigation in India* (Delhi: Mittal Publications, 1988), pp.290-91, 304.

[24] *Ibid.*, p. 303.

[25] *DA*, dec. VI (ii), p. 135.

[26] Correa, Gaspar, *Lendas da India* IV, (Lisboa: Academia Real das Sciencias de Lisboa, 1860), pp. 215, 289.

India wrote their diaries or log-books (*roteiros*) and drew maps with the help of locals, some completing them in India, others on their way or return to Portugal. In Goa, once Portuguese was enforced as the official language and Konkani banned, no need was felt of translations.

It is only with the introduction of paper during the 16th century that India was able to forge ahead in the progress of knowledge. But for this too, she had to rely on Portugal and other European countries for quite some time.